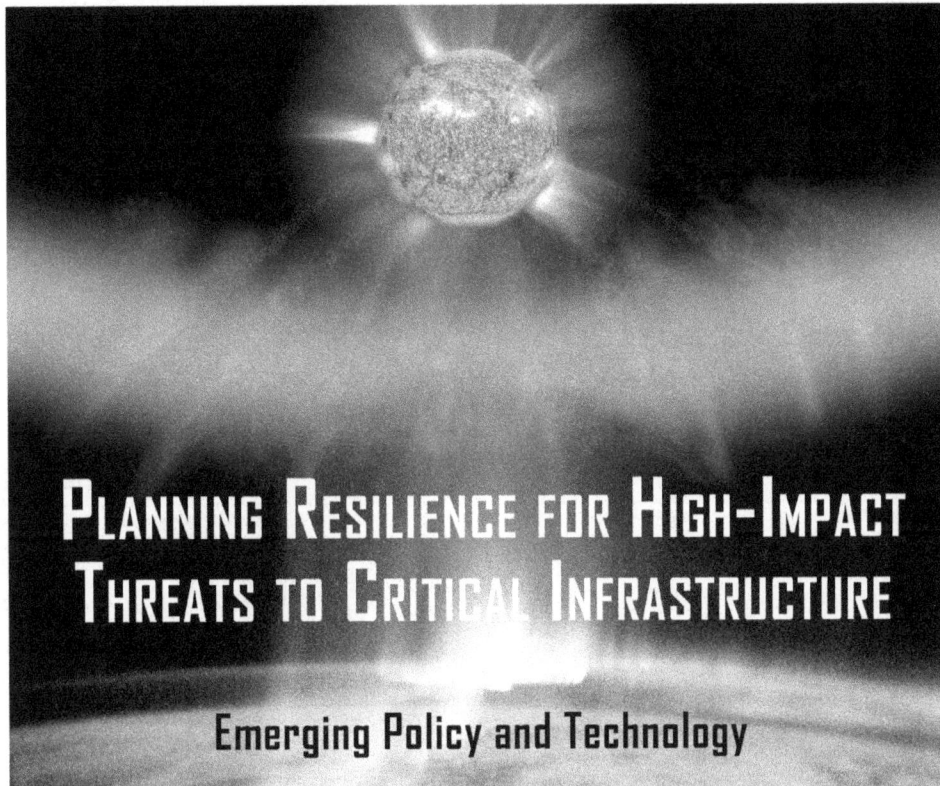

PLANNING RESILIENCE FOR HIGH-IMPACT THREATS TO CRITICAL INFRASTRUCTURE

Emerging Policy and Technology

CONFERENCE PROCEEDINGS OF THE
INFRAGARD NATIONAL EMP SIG SESSIONS

AT THE

DUPONT SUMMIT 2014

VIDEO-RECORDED SESSIONS AT
HTTP://WWW.IPSONET.ORG/CONFERENCES/THE-DUPONT-SUMMIT/INFRAGARD-VIDEOS-2014

EDITED BY CHARLES L. MANTO AND STEPHANIE LOKMER

Friday, December 5, 2014
Whittemore House
1526 New Hampshire Ave, NW
Washington, DC

Westphalia Press
An imprint of Policy Studies Organization
1527 New Hampshire Ave., NW
Washington, D.C. 20036
info@ipsonet.org

ISBN-13: 978-1-63391-261-8
ISBN-10: 1633912612

Cover design by Jeffrey Barnes:
www.jbarnesdesign.com

Daniel Gutierrez-Sandoval, Executive Director
PSO and Westphalia Press

Updated material and comments on this edition
can be found at the Westphalia Press website:
www.westphaliapress.org

Planning Resilience for High-Impact Threats to Critical Infrastructure

Conference Proceedings
InfraGard National EMP SIG Sessions at the 2014 Dupont Summit
December 5, 2014

At the Whittemore House
1526 New Hampshire, NW, Washington, DC 20005

Hosted and Published by the Policy Studies Organization

Edited by Charles L. Manto and Stephanie Lokmer

October 2015

© by Charles L. Manto and InfraGard EMP SIG

Book cover and interior design
by Jeffrey Barnes
jbarnesbook.design

Westphalia Press
An imprint of Policy Studies Organization
1527 New Hampshire Ave., NW
Washington, DC. 20036
info@ipsonet.org

Table of Contents

Preface and Acknowledgments

These conference proceedings of the InfraGard EMP SIG (Electromagnetic Pulse Special Interest Group) sessions at the Dupont Summit 2014 provide written presentations and background material for the video recordings available at http://www.ipsonet.org/conferences/the-dupont-summit/infragard-videos-2014.

The sessions covered high-impact threats to critical infrastructure with a special emphasis on geomagnetic disturbance (GMD), a topic of the sessions provided by the EMP SIG at each Dupont Summit since 2011 and the contingency planning workshops and exercises with the National Defense University and the Maryland Emergency Management Agency in October 2011. On the preconference day of December 4, 2014, the EMP SIG held a workshop and tabletop exercise at the National Guard Association of the U.S. headquarters on the triple threat of space weather, EMP, and cyber attacks with a nationwide power outage scenario of either 1, 3, or 12 months.

Subsequent to the Summit, the White House Office of Science and Technology Policy published a National Space Weather Strategy that has been added to the conference proceedings along with other notable items such as a related request for proposals from the Defense Department in June 2015. This was a small business innovation research (SBIR) program request for strategies to protect defense critical infrastructure from the effects of long-term regional and national blackouts due to high-altitude nuclear burst EMP or drive-by directed energy weapons. Both of these documents were ground-breaking initiatives showing that local communities need to be more resilient since the time to rescue might not be four days, but, possibly 40 or 400 days.

The National Space Weather Strategy has six goals, the second of which is to have the whole-of-community" conduct planning and exercises focusing on long-term regional and national power outages. This is precisely what the EMP SIG has been doing and continues to expand through its regional activities of its regional EMP SIGs. The 2014 Summit continued to set the stage for local planning that will be developed further in the 2015 Summit. The significance of this will become more apparent as the strategy becomes a coordinated federal action plan late in 2015.

This is the third year that we have published conference proceedings. Each transcript of the prior year's presentation is hyperlinked to YouTube videos. Other essays and items in the bibliography section are also hyperlinked. This year, the EMP SIG is also publishing a workshop and tabletop exercise program on the triple threat of space weather, EMP, and cyber attacks so that local communities and businesses can better prepare for long-term power outages as long as a year. In turn, that work can lead to greater investment and development of sustainable local infrastructure. A one-page overview of the triple-threat training program is also contained within this year's conference proceedings with program procurement information. Material will be made available to National Guard units, the National Governors Association, the National Association of Counties, and local InfraGard chapters.

The EMP SIG wishes to thank the Policy Studies Organization (PSO) for its generous support of the conference and the assistance of many including Ms. Stephanie Lokmer who helped with editing. We also wish to thank the National Guard Association of the United States who provided use of their facilities for workshops and tabletop exercises. Of course, the EMP SIG, as a nationwide special interest group of InfraGard, appreciates the strong support of the national board and staff, local chapters across the country, its members across all the 50 states and territories, and the Federal Bureau of Investigation who provides significant support to its InfraGard program.

As the chairman of the InfraGard EMP SIG and its conference session organizer, I welcome you to contact me for more information about InfraGard's EMP SIG and ways to participate in future activities. For information on InfraGard and how to join, see www.infragard.org.

Charles Leo Manto (cmanto@stop-EMP.com)
EMP SIG Chairman, InfraGard National

Program Details for
Planning Resilience for High Impact Threats to
Critical Infrastructure in 2015

Historic Whittemore House, Dupont Circle
1526 New Hampshire Avenue, Washington DC
Main Auditorium Schedule
Friday, December 5, 2014

8:00–8:30 **Registration**

8:30–8:35 **Introduction**

      Mr. Charles (Chuck) Manto, InfraGard National EMP SIG Chairman, provides a brief overview of the 2014 EMP SIG Dupont Summit, the purpose of the InfraGard National EMP SIG and introduces Dr. Wallace Boston.

      Session Chair: *Dr. Wallace E. Boston, President and Chief Executive Officer,* American Public University System, comments on the background of Senator Ron Johnson and introduces the senator.

8:35–8:50 **Remarks: "A U.S. Senator's Perspective on the Importance of EMP and High-Impact Threat Planning and Protection".** Senator Ron Johnson provides remarks comparing the nature of severe economic threats over the next 30 years and the high-impact of EMP that could happen at any time.

      Senator Ron Johnson (R-WI), Ranking Member,
      Subcommittee on Financial and Contracting Oversight
      Senate Homeland Security and Governmental Affairs Committee

8:50–9:10 **Presentation: "InfraGard and High-Impact Threats"**

      FBI Unit Chief John Pi introducing
      Section Chief John Riggi, Section Chief John Riggi provides a program overview of the InfraGard program including the role of the EMP SIG and its unique contributions to sustainable local communities and critical infrastructure protection from high-impact threats.

9:10–9:50 **Panel: "Pre-Traumatic Stress Disorder and High-Impact Events, Maintaining Public Calm and Order"** examines reasons why discussions and planning for major disasters are difficult and often avoided. Examples are given including planning accomplished in Howard County, MD for the possible event of a small ground-burst nuclear weapon near the White House and how the public should be encouraged to shelter in place for

9

24–48 hours rather than try to evacuate through fall-out clouds. Research was shown that depict best practice and that responsible dissemination of this information and this type of information does not result in panic.

Honorable Dr. Roscoe Bartlett, former U.S. Congressman, MD
Ms. Mary D. Lasky—Johns Hopkins Applied Physics Laboratory
Dr. Richard M. Krieg—Krieg Group
Dr. Ben Sheppard—George Washington University
Ms. Jessica Wieder—Environmental Protection Agency (EPA)

9:55–10:25 **Panel: "Federal, State, and Utility Plans for Grid Protection"** reviews and compares the activities of the state of Maine and Virginia in light of slow progress at the federal level. Encouragement of volunteer activities among utilities was discussed through implementation of the EIS "E-Pro Handbook".

Honorable Roscoe Bartlett, former U.S. Congressman
Dr. Chris Beck, Vice President, Electric Infrastructure Security Council
Senator Bryce Reeves, Virginia Commonwealth Senate
Honorable Andrea Boland, Outgoing Maine State Representative

10:30–10:55 **Panel: "Updates on Space Weather Threats for Power and Communications, Why 2012 Storm Became News in 2014"** Mr. Bob Rutledge reviews the latest findings from review of space weather data including the near miss of the super solar storm of 2012 and plans of the federal government in response. Mr. Nordling covers private sector investment response for critical infrastructure mitigation of space weather threats to electric power infrastructure and latest test results of the solutions offered by his company and used by a utility in the Midwest.

Mr. Robert Rutledge, Lead of NOAA Space Weather Forecast Office
Mr. Gale Nordling, CEO, Emprimus

11:00–11:20 **Presentation: "Dispelling the Myths about EMP"**

Mr. Curtis Birnbach, President, Advanced Fusion Systems discusses his company's approach to grid protection against manmade EMP and GMD from space weather. He also expresses his concerns about MIL SPECs for EMP that are later addressed by Dr. George Baker at 4:15.

11:25–12:10 **Panel: "Role of DHS Programs for EMP Protected Emergency Communications and Planning"** covers the work within the U.S. Department of Homelands Security to create EMP protection measures for emergency communications systems. Mr. Caruso explained work of the private sector to provide mitigation solutions. Mr. Bron Cikotas, active in both DoD and DHS activities over four decades, was not able to attend due to serious illness.

Mr. Kevin Briggs, DHS Team Chief NCCIC
Dr. George Baker, Professor Emeritus, James Madison University
Mr. Bronius Cikotas, Former Division Chief, Defense Nuclear Agency
Mr. Michael Caruso, ETS-Lindgren, Director, Government
   & Specialty Business Development

12:10–12:15 **Cameo Presentations**

12:15–12:50 **Lunch Break Procedures and Working Groups Announcement (Meal provided on site)**

Ms. Mary Lasky, Program Manager, John Hopkins University
Applied Physics Laboratory

12:55–1:00 **Welcome Back/Cameo Presentations (NY Regional EMP SIG).** Professor Mel Lewis described the formation of the NE Regional EMP SIG starting in the NY area and their planning for a series of discussions, workshops, and tabletop exercises for high-impact events.

1:00–1:10 **Key Note Presentation: "Targeted, Prudent Investments Against EMP: Building Practical Resilience Strategies"** describes prudent and cost effective mitigation measures that the private sector could take using general examples of early adoption of some of these measures prior to establishment of formal regulations and standards.

Dr. Paul Stockton, former Assistant Secretary of Defense
For Homeland Defense

1:10–2:10 **Panel: "Role of EMS and Tabletop Exercises—Planning for EMP and High-Impact Disasters"** reviews the planning and training of federal, state, and local government entities for high-impact disasters.

Dr. Paul Stockton, Panel Moderator, former Assistant Secretary of Defense
Dr. Richard Andres, Professor, National War College
Mr. Thomas MacLellan, National Governors Association
MG (ret'd) Robert Newman, U.S. AirForce, retired and former
Adjutant General of Virginia
Ms. Cynthia Ayers, former NSA Visiting Professor to the U.S.
Army War College (2003–2011)
Mr. David Hunt, Workshop and Exercise Lead Facilitator
Mr. Dennis Schrader, President, DRS International

2:15–2:55 **Panel: "Resilient Hospitals in Large-Scale Disasters (The Role of Alternative Technologies and Sustainability in Electric Power Grid Mitigation)"** compares the work underway in the U.S. DoD, the private sector and efforts in third world countries for hospital operation without the benefit of power grids.

Dr. James Terbush, Martin Blanck and Associates, Public
Health, Colorado Springs
Dr. Terry Donat, Independent Biosecurity Consultant an
IEMA RACES Officer, Chicago
Dr. Donald Donahue, Consultant to American Academy of
Disaster Medicine, Washington DC
Ms. Sierra Bainbridge, MASS Design Group, Boston
(Off-grid hospitals)
Mr. Art Glynn, CAPT USN Navy Emergency Preparedness
Liaison Officer, USNORTHCOM

3:00–3:40   **Presentation: "Growing Inter-Dependency of Gas and Electric Grids"** provides an overview of the growing interdependence of both gas and electric supplies providing a backdrop of information on growing vulnerabilities.

        Commissioner Philip D. Moeller, Federal Energy Regulatory Commission

3:40–4:15   **Panel: "Next Steps for States and Local Communities in Light of Vulnerabilities and Limited Federal Remedies:"** covers the efforts at the state level given slowness of federal level regulation and guidance to meet higher-impact threats.

        Mr. Bill Harris, International Lawyer and Secretary,
          Foundation for Resilient Societies
        Mr. Tom Popik, Chairman, Foundation for Resilient Societies
        Honorable Ms. Andrea Boland, former Maine State Representative
        Ambassador Hank Cooper, Chairman, High Frontier

4:15–4:25   **Presentation: "One More EMP Knot to Untie"** presents material that explains the role of MIL SPEC 188.125 and related testing of the U.S. DoD establishing its usage to protect defense department infrastructure. This counters a challenge brought earlier in the day by speaker Curtis Birnbach.

        Dr. George Baker, Professor Emeritus, James Madison
          University

4:30–4:55   **Panel Cameo Presentations (Solutions Update by sponsors and others)**
        Advanced Fusion
        Avaya
        Cyber Innovation Labs
        Distributed Sun
        Emprimus
        ETS-Lindgren
        JCTF
        MASS
        UET
        US DHS

5:00–5:30   **Economic Impact Studies, Next Steps for EMP SIG Working Groups and Concluding Remarks.** This segment provided an overview of next steps with emerging regional EMP SIGs and the workshops and tabletop exercises expected over the next year. It also provided a reminder of the EMP SIG meeting in April at the Space Weather Workshop (April 15) in Boulder, CO.

        Mr. Chuck Manto
        FBI Coordinators

Parallel Sessions at the Policy Studies Organizations HQ Garfield Room
1527 NEW HAMPSHIRE AVENUE, NW, WASHINGTON, DC 20036

4:20–5:25 pm (this section will not be webcast)

> Ms. **Denisa Scott.** Emerging mobile application innovations
> For mitigating/responding to threats to our electric grid and
> advancing community resiliency and recovery
>
> Ms. **Elysa Jones.** CAP Innovations/Services applicable to
> mitigating/responding to threats to our electric grid and
> advancing community resiliency/recovery
>
> Mr. **Joel Coulter.** Introducing wireless, surveillance, and
> affordable UAS remote sensing
> platforms strengthening communities critical infrastructure
> protection capacities.
>
> **Dr. David Bither**. An Orientation: Forward Trace's service
> oriented horizontal information
> exchange enterprise architecture's ability to optimize
> the performance of communities security investments, and
> proactive response to physical/cyber threats.

Registrants are also invited to a special lecture on Friday Night December 5 by the Policy Studies Organization:

Evening Session: 8:15 pm, as part of a Lecture Series of the Philosophical Society of Washington

John Wesley Powell Auditorium

The Purpose

The purpose of the InfraGard National EMP (electromagnetic pulse) SIG (special interest group) is to foster communications and coordination that will address and mitigate the threat of a simultaneous nationwide collapse of infrastructure from any hazard such as manmade or natural EMP.

Method Focusing on Local Sustainability

The National EMP SIG will mobilize subject matter experts at the national level so that local InfraGard chapters can make use of them to help local communities become more sustainable in light of these threats.

Resources

Expert National Advisory Panels

The National EMP SIG will establish panels of leading advisors. Examples may include but are not limited to

1. a **civilian-military liaison panel** enabling local communities to become more resilient so they can better support their local military and National Guard resources;
2. a **legislative and policy liaison panel** that can facilitate discussions between interested leaders at the national and local level with those in the private sector to identify and fill policy gaps;
3. an **education panel** that can facilitate research, development and education/training into the development of human resources needed;
4. an **investment panel** that might identify the capital support needed by local communities to enhance its sustainability;
5. a **media, communications and outreach panel** that can facilitate communications among EMP SIG members and between other stakeholders outside the SIG;
6. an **internal coordination panel** that would coordinate activities between the EMP SIG and other SIGs and committees within InfraGard; and
7. a **communications liaison panel** that would share information about emerging communications technology and make use of it to further the activities of the EMP SIG and InfraGard.

Qualifications and Expectations of National Panel Members

Qualifications: Panel members will be chosen based on their leadership within their respective fields by virtue of knowledge, experience, roles, capabilities, or relationships.

Expectations: Panel members will agree to attend an in-person meeting once per year, several conference calls during the course of the year and occasional email or phone correspondence with EMP SIG leadership. However, given the volunteer nature of these roles, it is expected that their contributions while meaningful will be limited and offered on a best-efforts basis.

Appointment of National Advisory Panel Members: Membership to the national panels will be by appointment by the EMP SIG Chair (manager). The initial chairman of each panel, if needed, will be appointed by the EMP SIG manager and subsequently by vote of the panel members at intervals of their selection.

The InfraGard National EMP SIG leadership team includes:

1. The InfraGard National EMP SIG Chair (manager) who serves at the pleasure of the InfraGard National Members Alliance (INMA) Board of Directors.

2. The InfraGard EMP Guidance Committee composed of the INMA Chairman, the INMA President, the INMA Managing Director, and an FBI HQ Supervisory Special Agent from the National Industry Partnership Unit (NIPU) who functions as an official liaison for the FBI to InfraGard on EMP matters.

3. The InfraGard EMP SIG Senior Management Team who is appointed by the EMP SIG Chair to serve the EMP SIG membership through the following positions:

   a. the Vice-chair who is familiar with key EMP SIG activities and can take the place of the Chair during periods of the Chair's unavailability;

b. the Administrative Officer, whose role is that of records keeper and other administrative functions of the Senior Management Team;

c. the Strategy Officer, who will help develop and align EMP SIG policy with its activities;

d. the Finance Officer, who will help track financial activities to support the EMP SIG and its coordination with the INMA treasurer;

e. the Regional Outreach Facilitator, who will assist the Senior Management Team and its activities with local InfraGard chapters and Regional Reps;

f. the Liaison Panels Facilitator, who will support the work of the Liaison Panels and their interactions with the Senior Management Team;

g. the Subject Matter Expert Panels Facilitator, who will support the work of the SME Panels and their interaction with the Senior Management Team; and,

h. other ad hoc committees that the Senior Management Team may deem useful from time to time.

EMP SIG Membership

Any InfraGard member in good standing may join the EMP SIG at the national level by indicating interest in participating in activities, mailings, or communications designed for membership participation. Members may either be designated as "working members" or "observers." EMP SIG members are encouraged to ask EMP SIG national leadership for help with their local EMP SIG activities.

Secure Communications between SIG Members

The InfraGard secure portal will be the primary means of communications between SIG members. EMP SIG leadership will also assist in providing other resources, including those that may be less secure, to supplement the official InfraGard communications and resources as needed on a best-efforts basis. This will include library resources and links to resources deemed to be of special value by the membership of the EMP SIG.

Other Resources

The EMP SIG will be responsible for recruiting and raising resources necessary to perform its tasks subject to the normal and customary procedures and governance of the InfraGard National Members Alliance and InfraGard Members Alliances.

Governance and Activities of the EMP SIG: All activities of the EMP SIG will comply with the governance and ethics as required by the INMA Bylaws and any guidance provided by the INMA Board of Directors.

(Background: The concepts of this guidance document have been proposed by the founding SIG Chair (manager) and approved by the INMA and the FBI NIPU. See the initial authorizing letter for the EMP SIG and the initial EMP Committee/SIG proposal for additional background.)

EMP SIG ((●))
ElectroMagnetic Pulse - Special Interest Group

Chuck Manto
EMP-SIG
Manager

EMP-SIG Senior Management Team

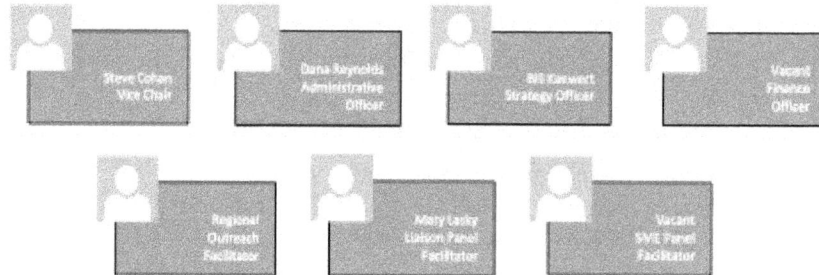

Steve Cohan
Vice Chair

Dana Reynolds
Administrative
Officer

Bill Kaewert
Strategy Officer

Vacant
Finance
Officer

Regional
Outreach
Facilitator

Mary Lasky
Liaison Panel
Facilitator

Vacant
SME Panel
Facilitator

EMP-SIG Subject Matter & Liaison Panels
INITIAL PANEL SELECTION APPROVED BY EMP-SIG MANAGER

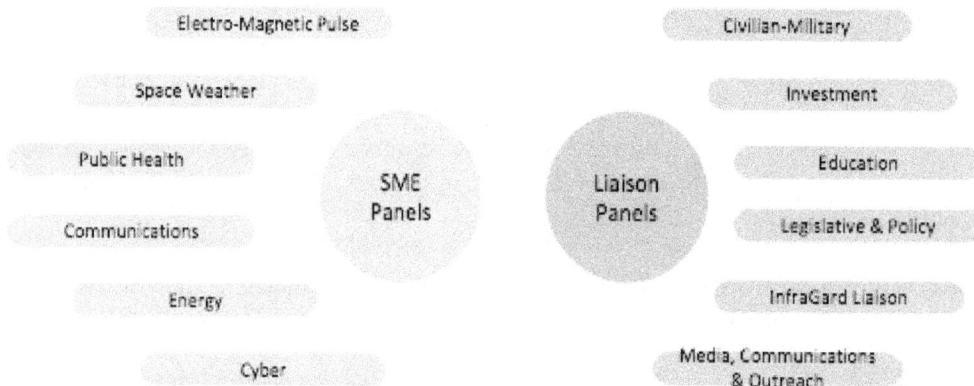

SME Panels
- Electro-Magnetic Pulse
- Space Weather
- Public Health
- Communications
- Energy
- Cyber

Liaison Panels
- Civilian-Military
- Investment
- Education
- Legislative & Policy
- InfraGard Liaison
- Media, Communications & Outreach

Proposed Organizational Structure

InfraGard Southeast Electromagnetic Pulse Special Interest Group

Charter Document

Scope

This document describes the initial working chartered, **InfraGard Southeast Electromagnetic Pulse Special Interest Group** herein referred to as **IES-SE** or **Regional SIG**.

This Regional SIG is a subset of the EMP SIG as chartered by the InfraGard National Members Alliance (*INMA*). The purpose of this regional entity is to coordinate, at a regional level, activities that are within the charter of the EMP SIG (*see National EMP SIG Guidance Document*). The need for this regional entity is driven, in large part, by the nature of the "low frequency, high consequence" solar, man-made nuclear, and man-made non-nuclear EMP threats to infrastructure, critical and otherwise, in order to facilitate information sharing and efforts that promote preparedness. The name is by intent interim and subject to change. It shall become the permanent name of this committee after one year of the approval of this charter if no other name is selected.

The EMP SIG mission is to address any high impact threat that could cause long-term nationwide collapse of critical infrastructure. These threats include but are not limited to intentional-EMP (nuclear and otherwise), extreme space weather, cyber-attacks, coordinated physical (kinetic) attacks, and widespread pandemics. The SIG's main focus is on EMP. The SIG serves to provide increased awareness to and between industry and business, academia, law enforcement and government regarding the potential threats and vulnerabilities that would have widespread and long lasting effects. The official documents of the EMP SIG such as its Guidance Document are posted on the EMP SIG sections of the InfraGard website at InfraGard.org.

Regional SIGs are intended to be a subsection of the InfraGard EMP SIG and, as such, is under the guidance and direction of the EMP SIG national management team and governance structure, specifically, it is not a separate entity but rather an extension of the national EMP SIG. The Regional SIG holds authority only in what is granted to it by the national EMP SIG and INMA, the basis for which is set forth in this document.

In no event are National EMP SIG or Regional SIG activities to imply authority over any InfraGard chapter and their activities. Chapter level participation and chapter level EMP SIG organization membership at a regional level is on a strictly voluntary basis.

Where this document may contain gaps in specificity, it will fall to the National EMP SIG first and subsequently INMA bylaws and policy to govern appropriate action and decisions.

The Regional SIG Executive Committee reports directly to the designated EMP SIG Regional Outreach Facilitator (national position). If the facilitator position is vacant the committee will report to the national EMP SIG chair through a designated interim facilitator. The Regional SIG will also work with the EMP SIG Liaison Panel as appropriate.

The Regional SIG will function as a provisional organization and will exist at a provisional level until such time as the National EMP SIG determines it appropriate to re-assign the Regional SIG from provisional to permanent status. Until the Regional SIG reaches permanent status, the National EMP SIG will be solely responsible for appointment of Executive Committee member positions.

This charter shall be reviewed at least annually by the Regional SIG Executive Committee and adjusted as necessary. Such revisions shall be submitted to the National EMP SIG chair for input and guidance. Any conflicts or exceptions to any policy are documented in the exceptions section of this document. **Regional SIG plans and structure require final approval from National EMP SIG Chair (*the requirements for this approval are set forth by the National EMP SIG*).**

Background

It is recognized that the EMP topic (sources and vulnerabilities) has special relevance to the energy and communications sectors because of the dependence of all of the other critical infrastructures (water, waste treatment, natural gas distribution, food supply, transportation, healthcare, law enforcement, etc.) on a steady supply of electricity and communications. The energy industry is comprised of a combination of local and regional assets. An important function of the National EMP SIG is to spread the word (to industry representatives, emergency management, elected government representatives, and to a degree—the public), regarding the vulnerabilities of the energy infrastructure (in particular) to an EMP and the long-term consequences of a widespread power outage. The ultimate goal of promoting information sharing between these stakeholders is to increase preparedness. The regional committee's purpose is to coordinate and implement that objective (and others of the EMP SIG) at the regional level and among their local communities.

Structure and Governance

At the 2014 National EMP SIG Summit, and through other activities also happening at regional levels, it became apparent that there is a need and desire to focus on these topics at a regional level. The intent of this is to help facilitate information flow and provide better resources to local InfraGard chapters and other emergency management service organizations.

The members of the Regional SIG share information regarding events, presentations and presenters, circulate documents and literature (both down to local and up to national levels), and focus on the exchange of EMP-related information for the greater good. All members shall be required to participate in the Oath of Office contained in *Appendix D*.

The region covered by the Regional SIG is modeled after the InfraGard regional structure which covers InfraGard member chapters within the specified region on a state-by-state level (*see Appendix A for states and member chapters covered by this Regional EMP SIG*).

The Regional SIG will identify opportunities to initiate, nurture, and refine capabilities that best support its mission in a manageable manner. These areas include the establishment of a governance framework, knowledge management, volunteer management, interaction with governmental and commercial entities, and opportunities to best plan, practice, and scale effectively. An integral component of this will be creating and managing exercises, and the capabilities to support such exercises, that promote the ability to respond and recover from threats such as HEMP. The Regional SIG will follow an iterative process of growth and sustainability activities, and adhere to the concept of "Think Big, Start Small, and Scale Fast".

The Regional SIG will be made up of active members in good standing of InfraGard chapters within the defined regional area. Should a member no longer be in active or good standing with InfraGard, they may no longer be eligible for membership with Regional SIG. Regional SIG will consist of an Executive Committee, Steering Committee, Oversight and Advisory Committee, and General Membership. The purpose of this structure is to provide regional guidance that is consistent with national goals and objectives as defined by the National EMP SIG but applied to the region. The membership base will be composed of interested InfraGard members within the region and will be responsible for, under guidance of the steering committee, with approval from the executive committee, the formation of working groups at the local and regional level to discuss, work through, and provide solutions to problems that are a focus of the EMP SIG as chartered by the Regional SIG. This includes issues from local to national levels as designated.

The Regional SIG executive committee has the ability to appoint advisors to the oversight and advisory committee as desired. The advisors have no voting rights on business but serve functions designated to them by the committee.

See *Appendix B* for the initial list of committee members and their roles.

The Executive Committee will transition to elected positions no sooner than 2018 or as deemed appropriate by the National EMP SIG. In the event that an executive committee position opens, the Executive Committee will be responsible for nominating a person for the position to be approved by the National EMP SIG. Should committee membership require members with federal security clearances, and in the event cleared replacement members cannot readily be assigned, existing cleared members with expired tenure may be extended on a year-by-year basis upon approval from the FBI and the INMA board, usually through the National EMP SIG Chair. *(As in the case of the National EMP SIG leadership, all Regional SIG leaders also serve at the appointment of the FBI and the INMA board, usually through the National EMP SIG Chair.)*

Executive Committee members shall be nominated out of the steering committee and subjected to approval of the National EMP SIG.

Executive Committee members shall serve two-year terms, not to exceed two consecutive two-year terms, re-election or appointment is allowed after a one-year wait period. The Executive Committee will determine the creation of any additional Executive Committee positions by simple majority.

To facilitate a smoother transition, unless no Executive Committee members are available, the Executive Committee may not be fully vacated; this is to allow new Executive Committee members to work with existing Executive Committee members that have institutional knowledge of process, procedure, relationships and other information that will be necessary for the new Executive Committee members to function effectively. Given this, no less than two veteran Executive Committee members shall be on the Executive Committee. In the event that all Executive Committee positions become vacant at the same time, all existing Executive Committee members will be placed on the ballot in a separate section, the two Executive Committee members with the highest votes will continue as active Executive Committee members at the conclusion of the voting. In this situation exclusively, term limits do not apply.

The Executive Committee can appoint advisory members as they see fit, Executive Committee term limits do not apply to these appointments, and are in effect for the term of one year with the ability to renew by Executive Committee vote.

All committee members and advisory members must be approved by the FBI and the INMA board, usually through the National EMP SIG Chair.

Steering Committee membership is comprised of appointees from local chapters within the region covered by the Regional SIG. Chapters are allowed to, but not required to, appoint one representative on the Steering Committee of the Regional SIG. Membership is limited to one formal position per chapter and limited to local chapter board members, sector chiefs and the leader of local EMP SIGs (*it is advised, but not required, that any local EMP SIG heads should be at least non-voting local chapter board members as well*). These memberships will remain in effect until the corresponding local InfraGard chapter determines otherwise.

General Membership is open to InfraGard members within the covered region in good standing for the purpose of working groups participation. This is considered open membership with no overall voting rights conveyed at the Regional SIG level.

General Membership oversight within the Regional SIG is the responsibility of the Executive and Steering Committees.

General Regional SIG Structure

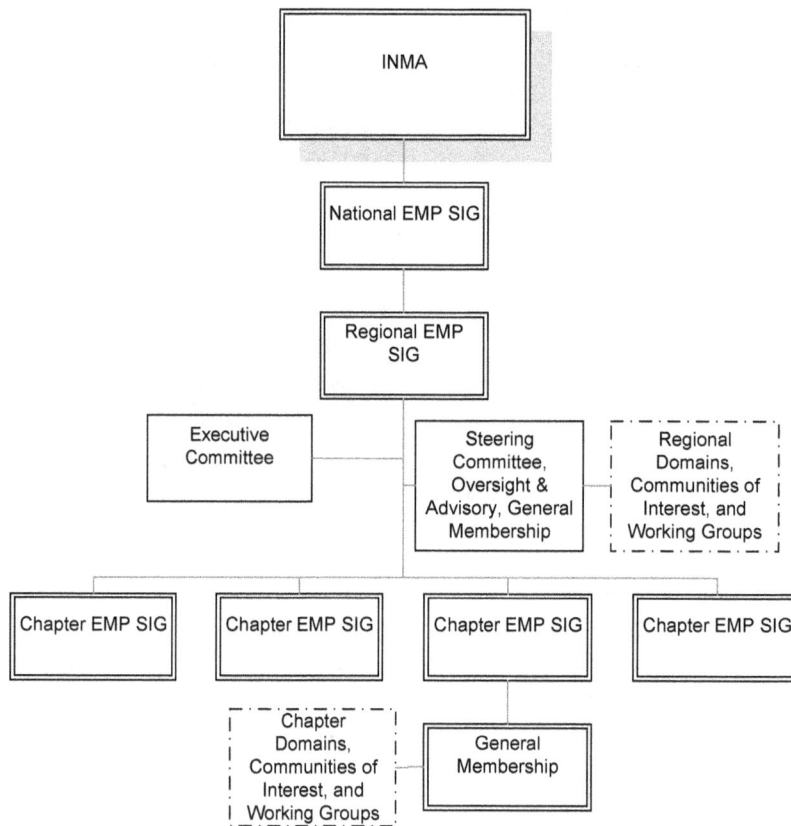

The above structure is, by intent, federated and designed to facilitate decentralized bottom-up problem solving with top-down centralized national and regional level oversight.

Purpose and Governance of Working Groups

A key part of enabling success is the ability to govern resources and workload within the Regional SIG. A main component of this will be the formation of task based working groups. The purpose of working groups is to organize various individuals to work on different tasks. These tasks can vary in nature based on the determined needs. For example, a working group can be formed to facilitate the creation of presentation material, to interface with energy organizations, to work on identifying possible solutions to a specific problem, and so on.

Working groups will be composed of interested members of the Regional SIG and can be proposed by any individual member of the Regional SIG. Each working group must be sponsored by a Steering Committee member and contain no less than three total members to be chartered. Working groups may contain or allow the limited participation of non-InfraGard and/or non-SIG members on a non-voting basis. Such participation must be sponsored in writing by an existing member and ensure that no access to confidential material at the "for official use only" or higher clearance level is compromised. Because the purpose of working groups is to facilitate development of ideas and execution of tasks, they may sometimes, but not always, align with different sectors within the Sector Chief's programs and, where such alignments take place, working groups would be expected to involve, cooperate, and work with SCs at the appropriate level.

Working Group Structure and Operation

*See **Appendix C**: Working Groups for additional details.*

Practical Application of Working Groups

There is no requirement or expectation for the entire Sector/Domain/COI/WG structure to exist unless there is a need for efforts in all areas covered. It is reasonable to expect only a limited portion of this structure to be active at any given time since the framework allows for a rapid turn-up/turn-down approach. Some structural flexibility will be the key to moving certain tasks and activities forward. These needs will be determined by the tasks at hand and utilize Working Group structure and operations as a guideline.

Communications

In order to facilitate effective and efficient communications, the Regional SIG Executive Committee Director will be the primary representative to the National EMP SIG regarding strategy, priorities, and planning. In the event that the Director is not available, the Deputy Director will be responsible for facilitating the communication. The purpose of this is to ensure a unified message both up and down the chain of the SIG.

The Executive Committee shall create and maintain a communications plan that addresses the cadence, content, participants, and back-up participants (per a succession plan) necessary to promote the Regional SIG mission. Additionally, the Regional SIG shall create and maintain an Operations Plan and other standard operating procedures that describe the organizational structure and processes for individuals operating under the direction of the Executive Committee Director. This Operations Plan shall include provisions for role-specific hierarchical communications and processes that facilitate governance and interaction between the various levels of the Regional SIG structure and external organizations. The Operations Plan shall include considerations for the Regional SIG involvement in both pre-event and post-event scenarios.

Members may serve as Regional SIG representatives or as a liaison to other organizations *only* as permitted in writing by the Executive Committee Director or Deputy Director. No member of the Regional SIG, to include Executive Committee Members, shall provide any statements to the media, press or via social media, or press releases that represent an official position of the Regional SIG, National SIG, INMA, InfraGard, or the FBI, without written authorization from the INMA board, usually through the National EMP SIG Chair, or their specific designee. Likewise, no member of the Regional SIG may officially represent the Regional SIG, National SIG, INMA, InfraGard, or the FBI in any official capacity unless authorized to do so in writing by the INMA board, usually through the National EMP SIG Chair, or their specific designee.

Funds, Sponsorship, and Contracts

The Regional SIG, being a subset of the National EMP SIG, will handle fund-related activities as a subset of the National EMP SIG. The Regional SIG will be incorporated into the National EMP SIG books as a subset of those finances. Local EMP SIGs will work with their local chapters to determine the best method for handling finances. This is intended to allow locally specific donations and funding at a state-by-state and city-by-city level to take place within the required scope of some funding requirements by more localized and state organizations.

The Regional SIG has the capacity to seek sponsorship to help offset costs associated with resources and activities only as allowed by the National EMP SIG to ensure that these activities are appropriately aligned. The Regional SIG cannot independently enter into contracts with any individuals or organizations except where specifically approved by the National EMP SIG. All contracts and fiduciary duties are the responsibility of the National EMP SIG or as delegated by the National EMP SIG to the Regional SIG Executive Committee, whichever is most appropriate and designated at the time of the action.

This may be subject to change with approval from the Regional SIG and National EMP SIG if appropriate to do so in the future.

Exceptions

See **Appendix F**.

Appendix A: Regional Memberships

The following states are within the designated InfraGard region and within the regional SIG scope. Participating chapters are listed beneath the state and updated as chapters participate.

- **North Carolina**
    - Western North Carolina (Charlotte)
    - Eastern North Carolina (Raleigh)
- **South Carolina**
- **Tennessee**
- **Georgia**
- **Alabama**
- **Mississippi**
- **Florida**
- **Puerto Rico (territory)**
- **St. Croix (territory)**
- **St. Thomas (territory)**

Appendix: B: Committees and Leadership

This appendix contains the initial leadership structure (*in alphabetical order*) for the SIG and will be updated as positions are filled.

Executive Committee

- Torry Crass, Western North Carolina (Director)
- Mike Hillier, Western North Carolina (Pending Confirmation)
- DJ O'Brien, Western North Carolina (Pending Confirmation)
- Stephen Volandt, Eastern North Carolina (Deputy Director)
- TBD

Steering Committee (Chapter Board Members/EMP SIG Chairs)

- Charlotte Chapter—Torry Crass
- Eastern Carolina Chapter—Stephen Volandt
- TBD
- TBD
- TBD
- TBD
- TBD
- TBD
- TBD
- TBD
- TBD
- TBD
- TBD
- TBD
- TBD

General Membership

- Mike Hillier—Charlotte Chapter
- DJ O'Brien—Charlotte Chapter
- TBD
- …

Oversight and Advisory

- [Permanent] National EMP SIG Chairperson—(Chuck Manto)
- [Permanent] InfraGard Chapter FBI Liaisons—All in Region
- [Appointed] InfraGard National Board Member—(Gary Gardner)

Appendix C: Working Groups

Working Groups shall be organized in a federated structure that facilitates local, regional and national alignment. Working groups will be organized into blocks called Sectors (in alignment with InfraGard) or Communities of Interest (COI). The Communities of interest will, at a minimum, represent the Federally designated critical infrastructure areas. Further, the COIs will be organized into Domains, which will report to the Executive Committee. The Regional SIG shall appoint all subordinate Domain and COI Chairpersons. The purpose of this structure is to facilitate governance, solution sharing, and the conflict resolution.

Working Groups provide the fundamental capabilities of the Regional SIG. The purpose of the Working Groups is to facilitate decentralized bottom-up solutions as part of a top-down centralized governance framework. The goal of all Working Groups is increased critical infrastructure resiliency during trigger events such as an EMP As stated in the charter, they shall at a minimum support the protection of Federally designated critical infrastructure sectors:

16 Federally Designated Critical Infrastructure Sectors and Sector-Specific Agencies

1. Chemical: Sector-Specific Agency: Department of Homeland Security
2. Commercial Facilities: Sector-Specific Agency: Department of Homeland Security
3. Communications: Sector-Specific Agency: Department of Homeland Security
4. Critical Manufacturing: Sector-Specific Agency: Department of Homeland Security
5. Dams: Sector-Specific Agency: Department of Homeland Security
6. Defense Industrial Base: Sector-Specific Agency: Department of Defense
7. Emergency Services: Sector-Specific Agency: Department of Homeland Security
8. Energy: Sector-Specific Agency: Department of Energy
9. Financial Services: Sector-Specific Agency: Department of the Treasury
10. Food and Agriculture: Co-Sector-Specific Agencies: U.S. Department of Agriculture and Department of Health and Human Services
11. Government Facilities: Co-Sector-Specific Agencies: Department of Homeland Security and General Services Administration
12. Healthcare and Public Health: Sector-Specific Agency: Department of Health and Human Services
13. Information Technology: Sector-Specific Agency: Department of Homeland Security
14. Nuclear Reactors, Materials, and Waste: Sector-Specific Agency: Department of Homeland Security
15. Transportation Systems: Co-Sector-Specific Agencies: Department of Homeland Security and Department of Transportation
16. Water and Wastewater Systems: Sector-Specific Agency: Environmental Protection Agency

http://www.whitehouse.gov/the-press-office/2013/02/12/presidential-policy-directive-critical-infrastructure-security-and-resil

Working groups shall be established at the regional, state, and substate levels. Likewise, these working groups shall be organized into Communities of Interest (COI) at their corresponding levels. These COIs shall be further organized into domains that facilitate cooperation and integration of related COIs.

Initially, the COIs will be organized into the below Domains. Domain structure will be determined at the discretion of the Regional SIG until a national consensus is determined by the National EMP SIG Chair, after which they will mirror the national structure. Domains and their respective COIs and Working Groups shall be established in a federated structure at the national, regional, state, and substate levels, and shall be aligned with their higher and lower Domains, COIs, and Working Groups by charter name and mission statement.

Population Critical Infrastructure Domain

1. Communications COI
2. Water and Wastewater (residential) COI
3. Food (residential) COI
4. Emergency Generator refueling COI
5. Transportation Systems COI
6. Emergency Services COI
7. Healthcare and Public Health (local) COI
8. Shelter/Housing COI
9. Energy (household) COI

General Government Domain

1. Communications COI
2. Nuclear Reactors, Materials, and Waste COI
3. Emergency Generator refueling COI
4. Energy (grid) COI
5. Water and Wastewater Systems (PUD) COI
6. Healthcare and Public Health COI
7. Public Safety/Law Enforcement COI
8. Government Facilities COI
9. Dams COI

Commercial Domain

1. Communications COI
2. Nuclear Reactors, Materials, and Waste COI
3. Emergency Generator refueling COI
4. Chemical Manufacturing COI
5. Information Technology COI
6. Financial Services COI
7. Food and Agriculture (commercial) COI
8. Commercial Facilities COI
9. Critical Manufacturing COI
10. Defense Industrial Base COI
11. Legal Services COI

Human/Social Critical Infrastructure Domain

1. Communications COI
2. Public Safety/Law Enforcement COI
3. Basic Education (3 Rs) COI
4. Constitutional/Moral Foundation of America Education COI
5. Economic Education COI

Note that certain communities of interest span more than one domain. Such communities of interest shall automatically be candidates for the Enterprise Services domain. It is expected that recovery from a no-notice EMP event could require 3–36 months depending upon its severity. During times of such emergency, Enterprise Services require a higher level of coordination and will be governed such that they directly support state and regional priorities. Working groups in such COIs will include consideration of this higher level of coordination requirement for a no-notice EMP event and the current capability to respond and recover from such an event. Additionally, working groups associated with Enterprise Services shall include consideration for mechanisms that allow centralized command and control as directed by each state governor or similar authority. Such consideration shall be their top priority.

Enterprise Services Domain (initial candidates)

1. Communications COI
2. Nuclear Reactors, Materials, and Waste COI
3. Emergency Generator refueling COI
4. Energy (grid) COI
5. Healthcare and Public Health COI
6. Public Safety/Law Enforcement COI
7. Water and Wastewater COI
8. Food and Agriculture COI
9. Transportation Systems COI.

As resiliency planning and preparations proceed over time, Enterprise Services working groups are expected to transition in and out of the Enterprise Services domain. Their Regional Steering Committee will determine such transitions. For example, the Communications COI may become adequately resilient such that centralized command and control is no longer required for its success, thus allowing its transition out of the Enterprise Services domain. Working Groups must consider this contingency as part of their planning. Any COI assigned to the Enterprise Services domain by any state-level governance committee shall automatically be assigned to the Enterprise Services domain at the regional level.

Post-Event Support

Many Working Group members will also be directly involved in post-event response and recovery as part of their regular non-Regional SIG job duties. The relationships and combined expertise established in certain Working Groups may prove an invaluable post-event combination. Working Groups and each COI, Domain, and Regional Steering Committee shall plan for the contingency of being of use during post-event response and recovery. This shall include consideration of involvement in post-event Enterprise Services governance, resilient communications, and the best means to support the life cycle of response and recovery activities.

Special Working Groups

The Regional Executive and Steering Committees may form Special Working Groups that directly report into the respective committee without the Domain/CoI structure beneath it. One example is to form Working Groups that provide staff-type services that may not be available via full time funded positions. Additionally, Special Working Groups may be created to span multiple geographies, or industry and technical domains similar to those the National EMP SIG designates as "Liaison Panels". Such groups shall be broad in scope, such as education, investment, media, faith-based and volunteer organizations, and policy.

Working Group Operations

Working groups shall be established at the discretion of their respective COI chairperson. The COI chairperson shall encourage the creation of working groups such that they facilitate effective problem solving and planning. COI chairpersons may nominate successful working groups for national, regional, state, and substate replication via their respective governance committees. COIs will be associated with one another in a Federated fashion such that the parent domain and COI structure is the same at the national, regional, state, and substate levels. Each working group shall be named such that it is easily associated with the overall national and regional working group structure. Each working group charter shall describe its alignment with the overall Federated and other working groups at higher and lower domain levels. This shall be accomplished by naming the aligning organizations and describing in its mission statement how it supports the larger structure's mission. For example, a state level emergency management communications working group would list its alignment with national, regional, and substate working groups by name, and describe how it supports the mission of these organizations.

Each Working Group shall be established via a charter that includes governance provisions consistent with higher level COI and Domain guidelines and approved in writing by the COI chairperson. Each charter will provide the Working Group Mission Statement, describe Working Group membership goals, name its location in the federated domain structure to include known associated higher and lower Working Groups, name the chairperson, meeting cadence, and methods of communications. Each Working Group charter shall name is officer roles (chair, deputy, secretary, etc.) and include provisions for record keeping and periodic reporting to their respective COI not less than monthly, quarterly, and annually regarding mission statement progress and alignment with the larger federated agenda. The term of the chairperson and other officers shall be for one year. The Working Group's founding officers shall be appointed by the COI, and selected by Working Group member's vote thereafter, with approval of the COI chair. The COI Chair may replace any Working Group officer at any time for any reason. Domain chairpersons may replace any COI officer at any time for any reason. All Domain and COI officers may be removed at any time by the FBI and the INMA board, and such action would be usually be accomplished through the EMP SIG Chair.

Working Group, COI, and Domain member's responsibilities follow:

Working Group Chairperson

1. Call and chair meetings.
2. Approve agendas.
3. Present and represent positions of the Working Group/COI/Domain.
4. Assign actions and tasks.
5. Convene and assign teams to perform specific tasks or develop specific products needed by the Working Group/COI/Domain.
6. Resolve other issues as required.
7. Seek Working Group/COI/Domain consensus on issues aligned with the Working Group/COI/Domain mission statement.

8. Approve minutes.

Working Group Members' Responsibilities

1. Identify and nominate agenda items and issues to the Chair for consideration.

2. Sponsor items and issues for meetings, including preparation of position papers and read-ahead materials, and presentation of briefings.

3. Represent their organizations' positions with regard to Working Group/COI/Domain issues.

4. Convey and support the positions and decisions of the Working Group/COI/Domain to their organizations.

5. Execute actions and tasks as directed by the Chair.

6. Ensure their organizations are represented on appropriate Working Group/COI/Domain subordinate bodies.

7. Keep the Working Group/COI/Domain apprised of relevant and significant matters aligned with the Working Group/COI/Domain mission statement.

8. Review minutes.

Secretary's Responsibilities

1. Propose issues and processes to support the functions of the aligned with the Working Group/COI/Domain mission statement. Conduct reviews of pertinent documentation, requirements, resource allocations, acquisitions, and waivers as directed by the National EMP SIG Chair or the Working Group/COI/Domain.

2. Provide advice and counsel to the Chair on Working Group/COI/Domain matters.

3. At the direction of the Chair, formulate, research, and present issues aligned with the Working Group/COI/Domain mission statement.

4. Structure issues and ensure proper representation on items before the Working Group/COI/Domain.

5. Announce and produce meetings at Chair's direction.

6. Ensure all security rules and regulations regarding classified/official-use-only meetings and documents are followed.

7. Assemble, prepare, and distribute material on matters under consideration by the Working Group/COI/Domain at least four working days in advance.

8. Disseminate specific requirements or recommendations for data and other actions on behalf of the Working Group/COI/Domain.

9. Disseminate to appropriate EMP SIG officials, decisions reached by the Working Group/COI/Domain.

10. Monitor and track follow-on actions taken to ensure that decisions reached and assignments made by the Chairperson are implemented properly.

11. Prepare and distribute minutes of Working Group/COI/Domain meetings.

12. Maintain and safeguard records and ensure their appropriate disposition.

13. Maintain liaison to the Working Group/COI/Domain bodies that are contained in higher and lower levels of the federated hierarchy. Maintain liaison with adjacent bodies as directed by the Chair, higher bodies, or otherwise as needed.

14. Support and coordinate the activities of the Working Group/COI/Domain subordinate bodies.

15. Compile and maintain contact lists for Working Group/COI/Domain members and their coordinating staffs.

Working Group chairpersons shall be appointed by their respective COI chairperson, and will serve at the discretion of the COI chairperson. Similarly, COI chairpersons shall be appointed by their respective Domain chairperson, and will serve at the discretion of the Domain chairperson. Domain chairpersons shall be appointed by their respective Committee chairperson, and will serve at the discretion of the Committee chairperson. *(As in the case of the Committee leadership, All Working Group, COI, and Domain leadership also serve at the discretion of, and may be removed by, the FBI and the INMA board, through the EMP SIG Management Team.)*

COIs and Domains shall be organized in a federated structure such that they report to and are governed by governance committees at their respective level in the hierarchy. For example, the substate working groups shall be members of COIs and Domains within that substate area, and shall be governed by a substate Committee, which in turn shall report to and be governed by a state level governance committee, which in turn reports to the Regional SIG committee. Each lower level governance committee must be approved by its respective parent level governance committee. Substate committees' quantity and territory shall be determined by the state level committees.

General State Structure

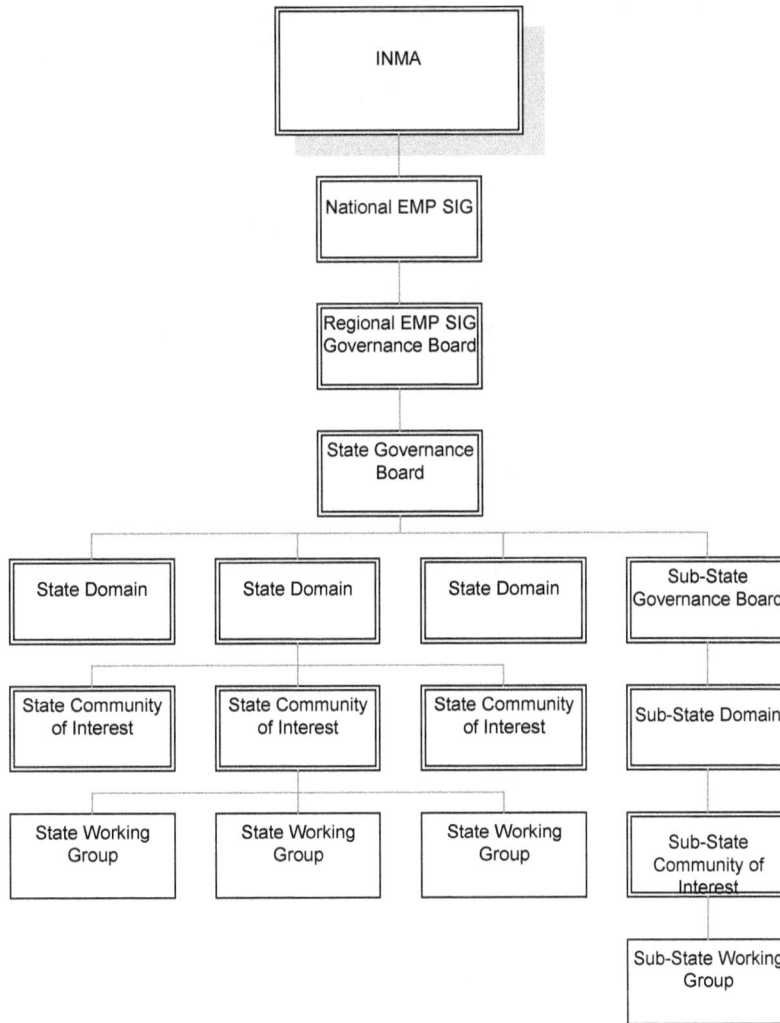

```
                    ┌──────────────────────┐
                    │         INMA         │
                    └──────────┬───────────┘
                               │
                    ┌──────────┴───────────┐
                    │  National EMP SIG    │
                    └──────────┬───────────┘
                               │
                    ┌──────────┴───────────┐
                    │  Regional EMP SIG    │
                    │  Governance Board    │
                    └──────────┬───────────┘
                               │
                    ┌──────────┴───────────┐
                    │  State Governance    │
                    │        Board         │
                    └──────────┬───────────┘
                               │
        ┌────────────┬─────────┴──────┬──────────────┐
   ┌─────────┐  ┌─────────┐    ┌─────────┐    ┌──────────────┐
   │  State  │  │  State  │    │  State  │    │  Sub-State   │
   │ Domain  │  │ Domain  │    │ Domain  │    │ Governance   │
   └────┬────┘  └────┬────┘    └────┬────┘    │    Board     │
        │            │              │         └──────┬───────┘
   ┌─────────┐  ┌─────────┐    ┌─────────┐    ┌──────────────┐
   │  State  │  │  State  │    │  State  │    │  Sub-State   │
   │Community│  │Community│    │Community│    │   Domain     │
   │ of Int. │  │ of Int. │    │ of Int. │    └──────┬───────┘
   └────┬────┘  └────┬────┘    └────┬────┘           │
   ┌─────────┐  ┌─────────┐    ┌─────────┐    ┌──────────────┐
   │  State  │  │  State  │    │  State  │    │  Sub-State   │
   │ Working │  │ Working │    │ Working │    │ Community of │
   │  Group  │  │  Group  │    │  Group  │    │   Interest   │
   └─────────┘  └─────────┘    └─────────┘    └──────┬───────┘
                                              ┌──────────────┐
                                              │ Sub-State    │
                                              │  Working     │
                                              │   Group      │
                                              └──────────────┘
```

Working Group Conflict/Idea Resolution

It is the Working Group's purpose to facilitate the healthy lifecycle of idea sharing and solution management. This will occasionally create differences of opinion that must be resolved. Each Working Group is chartered to pursue topics that further the National EMP SIG agenda. It is the responsibility of each COI Chairperson to ensure that Working Groups under their cognizance remain so aligned and to encourage the creation of Working Groups that stimulate problem solving and alternative solutions. Any member in good standing may nominate the creation of a Working Group at any level. The COI chairperson will provide written guidance to each of their Working Groups and monitor their progress via Working Group reporting, status meetings (at a reasonable tempo decided by the COI chairperson) and personal interaction as needed.

Working Group chairpersons shall serve as facilitators and strive to gain consensus. Working Groups shall use Robert's Rules of Order for meeting governance. All decisions shall be put to a vote with a simple majority needed for decision. The chairperson may overrule simple majority vote results, however, may be overturned by a 2/3 majority vote by the Working Group members. Either the Working Group chairperson or Working Group membership may escalate the decision to next higher level of governance for resolution if overruled.

When escalated, the higher authority shall ensure that the perspective of subject matter experts nominated by each opposing party is provided in the form of formal presentations. Such presentations shall be allowed a minimum of one hour if desired by the presenter, but no more than two hours. Participation in escalation proceedings may be virtual if practicable.

Either party may continue to escalate the decision to the next higher level of governance. The intent of voting and related escalation decision governance described herein is rapid decision making and shall be completed within one meeting cycle (at each governance level). Special meetings to accelerate the process may be called by the Working Group chairperson or a simple majority vote by Working Group membership, except when escalated. Escalated meetings shall be scheduled, within one month of receipt of the request, by the higher authority unless the escalating party declares it an urgent issue. In that case, the National EMP SIG chair (or designee) shall set the resolution meeting timing and venue. COI and Domain escalation governance shall follow this same construct as described above for Working Groups.

For issue resolution purposes, the COI shall serve as the level higher than the Working Group, and the Domain as the next higher level. Once at a Domain level, escalation shall proceed to it's respective governance committee (substate, state, or regional) and thereafter to the next higher governance committee in the federated hierarchy, with the ultimate decision level at the National EMP SIG level. Decisions shall be made at each level as described above for the Working Group. In the event of a tie vote, the respective Chairperson shall exercise overrule rights to make the final decision.

Escalation Path

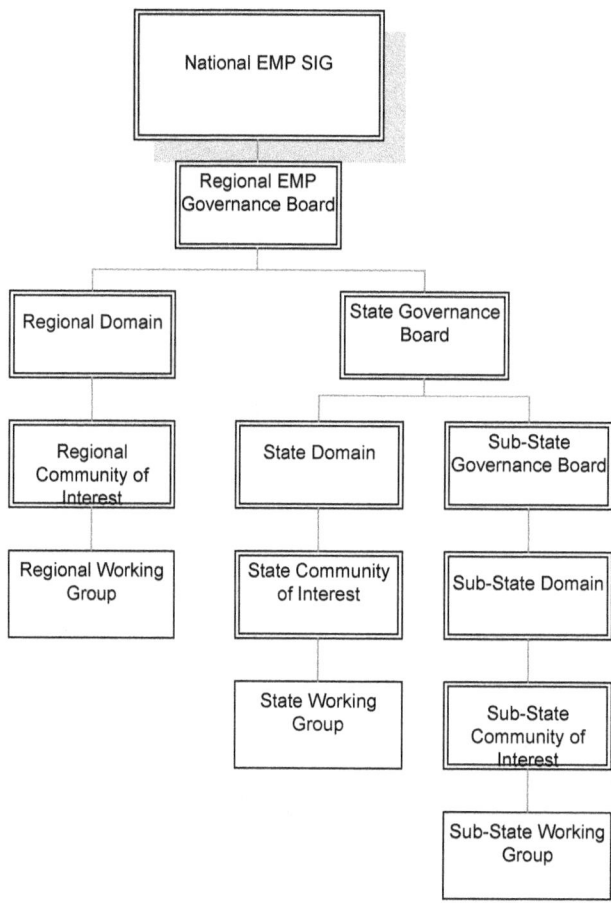

Appendix D: Oath of Office

The oath of office is expected for all Regional SIG members as an affirmation that they will execute their duties and responsibilities in good faith in the best interest of the country.

I, _____, do solemnly affirm that I will support and defend the Constitution of the United States against all enemies, foreign and domestic, that I will bear true faith and allegiance to do the same; so help me God.

Appendix E: Exceptions

In order to create a functional executive committee, the initial makeup of the Executive Committee will include multiple members from the same InfraGard chapters. This will remain in effect as long as those members wish to serve in their confirmed positions or until executive committee elections take place at which time those positions would need to transition to members of the Steering Committee that are appointed/elected to the Executive Committee.

No additional exceptions are currently in place.

Appendix F: Charter and Appendix To Do List

Update the COI—Community of Interest to Sectors to better align with the Sector Chief program.

Setup models for working groups, create small, medium, and large versions.

Add definitions to Appendix G where appropriate.

Appendix G: Definitions

SIG: Special Interest Group

Regional SIG: The regional entity related to the special interest group referenced.

National SIG: The national entity related to the special interest group referenced.

INMA: InfraGard National Members Alliance, the national governing body of InfraGard.

Sector: May also be called a community of interest. This is an area of specific focus related to critical infrastructure impacts as designated by the FBI. Currently there are 16.

Community of Interest: COI. *See sector…*

Working Group: A group of people formed through the SIG in order to facilitate and accomplish a task.

Executive Committee: Regional SIG leadership group comprised of officer positions.

Steering Committee: Chapter membership participants who help to facilitate the formation of Working Groups and address Regional SIG business.

Oversight and Advisory Committee: Individuals appointed by the Executive Committee to assist with designated tasks. Permanent members of this committee are the National EMP SIG chairperson and all FBI Liaison Officers.

General Membership: Non-voting members of the Regional SIG who contribute to activities by serving in Working Groups.

BACKGROUND

On December 4, 2014, the InfraGard Electromagnetic Pulse Special Interest Group (EMP SIG) conducted a workshop and tabletop exercise for about 200 people. The EMP SIG management team has been working to refine the documentation used so that others may be launched into holding their own exercises.

In addition, EMP SIG management team members conducted short exercises and provided a PowerPoint presentation others may use.

EMP SIG OFFERINGS

The EMP SIG is providing material for InfraGard members and others to conduct tabletop exercises that help participants think about and prepare for a breakdown of critical infrastructure. There are two different approaches that people may take; both have been vetted by subject matter experts. The two approaches are:

1. *1–3-Day Program:* Use the "High Impact Threats to the Electrical Grid Workshop and Tabletop Exercise" document with background material covering EMP, extreme space weather, and cyberattack scenarios of three varying impacts: 1-, 3-, or 12-month nationwide outages. This document is the result of the exercise held on December 4, 2014 in Washington DC, as a pre-event to the Dupont Summit. A facilitator guide is available to help a team conduct its exercise and to walk the planning committee through a detailed process or methodology. This approach outlines steps needed to: obtain a team; design objectives; develop roles and responsibilities; design the exercise; review and modify the scenario provided by the InfraGard EMP SIG; evaluate the exercise; etc.

2. *1-Hour Program:* Use the "Societal Effects of High Impact Threats to Critical Infrastructure" PowerPoint presentation. This is a simple approach that can be conducted in an hour or less, as a one-time event or as a starting point in building a team to develop the more in-depth exercise described in Approach #1. Approach #2 requires limited work to present background information or lead a discussion when using the annotated PowerPoint provided. The introductory slides may be used to provide the presenter's background, the group's perceived needs, and requirements. Qualified EMP SIG members are available to help facilitate the exercise should this be needed. For more information on facilitation assistance, send an email to: igempsig@infragardmembers.org

STEPS REQUIRED

1. Request the link to the materials at
   igempsig@infragardmembers.org

2. Use the material and conduct an exercise.

3. Complete the survey found at https://www.surveymonkey.com/s/YNYZYH5 to help the InfraGard EMP SIG management team continually improve exercise scenarios and documentation. Participant survey is at https://www. surveymonkey.com/s/Y5HK9MH

Please contact EMP SIG with any questions: igempsig@infragardmembers.org

The EMP SIG facilitates information sharing about threats leading to long-term failure of life-sustaining infrastructures and resources so that local communities can mitigate them and become more sustainable.

Biographies of Presenters

Anderson, Mr. George
Email: ganderson@crowniron.com

Mr. Anderson was originally employed at the Crown Iron Works Company, a family enterprise founded in 1878. There he served as Vice President for Engineering from 1975 until his departure. He also served in the capacity of Director and Executive Vice President where he specialized in new process development, design of major equipment, safety engineering, and international sales support. Mr. Anderson left Crown in 2009 and founded Emprimus LLC, a company devoted to the production of products that defeat the effects of geomagnetically induced currents and electromagnet interference caused by solar flares, RF weapons, and nuclear EMP. In addition to serving as Chairman of Emprimus, Mr. Anderson has been a 32-year member of the National Fire Protection Association and is an active member of the Committee on Solvent Extraction Plants (NFPA #36), which Mr. Anderson considers to be an excellent example of cooperation and balance between equipment manufacturers, plant engineers and operators, insurers, fire marshals, and various consultants in producing standards to reduce risk in very large systems handling flammable solvents in an extraction process—informed, flexible, performance-based, near-consensus standards accepted with little resistance or friction in many nations worldwide. Mr. Anderson has an Engineering degree from Stanford University, 1969. He also served as President of the Stanford *American Society of Mechanical Engineers* student organization, 1968.

Andres, Dr. Richard
Email: rich.andres@gc.ndu.edu

Dr. Andres is a Senior Fellow and Energy and Environment Security and Policy Chair at INSS of the National Defense University at Ft. McNair, Washington DC. His current work focuses on energy and environmental security and particularly defense-related energy issues. Prior to joining INSS, Dr. Andres was a professor at Air University assigned to the Pentagon where he served as Special Advisor to the Secretary of the Air Force. He has also served as a consultant to the Office of the Secretary of Defense (during both the Clinton and Bush administrations), the Joint Chiefs of Staff, the Office of Force Transformation, U.S. Strategic Command, the Nuclear Posture Review, the Council on Foreign Relations and other organizations. His publications appear in such journals as *International Security*, the *Journal of Strategic Studies, Security Studies*, and *Joint Force Quarterly*. Dr. Andres was awarded the medal for Meritorious Civilian Service, and has received numerous academic awards and fellowships. He received his PhD from the University of California, Davis.

Sample Articles: "Volatility in the European Energy Security Framework: Addressing Ukraine–Russia Gas Pricing Disputes," INSS Strategic Forum (2nd Quarter, 2010). "Small Nuclear Reactors for Military Installations: Capabilities, Costs, and Technological Implications," INSS Strategic Forum (2nd Quarter, 2010).

"Energy and Environmental Insecurity," Joint Forces Quarterly, Vol. 55 (4th Quarter, 2009). "The Emerging Energy Security System," book chapter, in Global Strategic Assessment (National Defense University Press, 2009). "The Department of Defense: New Energy Infrastructure and Fuels," book chapter, in Global Strategic Assessment (National Defense University Press, 2009).

Ayers, Ms. Cynthia
Email: cayers@tfnhs.comcastbiz.net

Cynthia Ayers is currently Deputy to the Executive Director of the Task Force on National and Homeland Security. Ms. Ayers retired from the National Security Agency (NSA) in 2011 and was subsequently appointed as Vice President of EMPact America.

Her intelligence community career included a position as an NSA Representative to the DCI's Counterterrorism Center, where she worked throughout the attacks on the USS Cole and the attacks of 9/11 (2000–2002). She served at the Center for Strategic Leadership as the National Security Agency's Visiting Professor to the U.S. Army War College (USAWC). While at the USAWC, Professor Ayers taught on contemporary threats to national security from an intelligence perspective, and advised students on research concerning strategic intelligence, counterterrorism, cyberwarfare, the Middle East, and critical infrastructure protection.

Ms. Ayers is also a consultant at the Center for Strategic Leadership (USAWC), a Fellow with the Center for Advanced Defense Studies, a member of the Secure the Grid Coalition and a member of the FBI InfraGard EMP Special Interest Group (SIG), as well as Vice President of the Central PA Chapter of InfraGard.

Bainbridge, Ms. Sierra
Email: sierra@mass-group.org

Sierra Bainbridge is the senior director at MASS Design Group. Sierra began work with MASS in 2008 focusing on the landscape architecture, and joined full time in 2009 to finalize design and oversee implementation of the Butaro Hospital, MASS's first project. Currently Sierra directs the ongoing design and implementation of MASS's planning and architectural projects and is currently overseeing The Kayanja Center, an academic facility supporting rural healthcare delivery and research in Uganda, a number of African Conservation Schools in DRC, Tanzania, Zambia, and Rwanda, and the Butaro Hospital Expansion Plan, among others. Those completed include Butaro Hospital, the Umubano Primary School, the Butaro Doctors' Housing, and the Butaro Ambulatory Cancer Center.

Prior to joining MASS, Sierra worked for four years at James Corner Field Operations, primarily in design and oversight of implementation of Section 1 of the New York City High Line. Sierra has taught graduate level studios at various universities and from 2010 to 2012, Sierra served as Head of the Architecture Department at the Kigali Institute of Science and Technology (KIST) in Rwanda. At KIST, Sierra was instrumental in shaping the current curriculum. She is invited to speak regularly, including the keynote address at the Healthcare Design Conference, serving as a Sasaki Distinguished Visiting Critic at the Boston Architectural College, and lecturing at the Carter "Lectures In African Studies" series, the University of Pennsylvania, Harvard University, University of Toronto, and the American Institute of Architects, among others. Select features of Sierra's work with MASS Design Group include A+U Magazine, Lotus, Mark Magazine, and Detail. Sierra received her Bachelors of Arts in Art and Architectural History from Smith College and her Masters of Landscape Architecture and Masters of Architecture from the University of Pennsylvania.

Baker, Dr. George
Email: George Baker_CONTRACTOR@dtra.mil

Dr. Baker is emeritus professor of applied science at James Madison University (JMU). In addition to teaching graduate and undergraduate S&T courses at JMU, he directed the start-up and served as Technical Director of the university's Institute for Infrastructure and Information Assurance (IIIA). Much of his career was spent at the Defense Nuclear Agency (DNA) and the Defense Threat Reduction Agency (DTRA) protecting strategic systems against electromagnetic pulse (EMP) and developing protection guidelines and standards.

He led DNA's EMP research and development program during 1987–1994 and recently served as principal staff for the Congressional EMP Commission. A primary research interest stems from his experience as Director, Springfield Research Facility—a national center for critical system vulnerability assessment. He applies lessons-learned from DoD experience to critical national infrastructure assurance and community resilience. He consults in the areas of critical infrastructure protection, EMP and geomagnetic disturbance (GMD) protection, nuclear and directed energy weapon effects, and risk assessment. He presently serves on the Board of Directors of the Foundation for Resilient Societies, the Board of Advisors for the Congressional Task Force on National and Homeland Security, the JMU Research and Public Service Advisory Board, and the National Defense Industrial Association (NDIA) Homeland Security Executive Board.

Bartlett, Congressman Roscoe G.
Email: roscoegbartlett@gmail.com

Elected to serve his 10th term in the U.S. House of Representatives, Roscoe G. Bartlett considers himself a citizen-legislator, not a politician. Prior to his election to Congress, he pursued successful careers as a professor, research scientist and inventor, small business owner, and farmer. He was first elected in 1992 to represent Maryland's Sixth District.

In the 112th Congress, Bartlett serves as Chairman of the Tactical Air and Land Forces Subcommittee of the House Armed Services Committee. Owing to his 10 years of experience as a small business owner, he also serves on the Small Business Committee. One of three scientists in the Congress, Dr. Bartlett is also a senior member of the Science, Space, and Technology Committee.

Prior to his election to the Congress, Dr. Bartlett worked for more than 20 years as a scientist and engineer on research and development programs for the military and NASA. Nineteen of his 20 patents are held by the U.S. Government for his inventions of life support equipment used by military pilots, astronauts, search and rescue personnel, and firefighters.

In 2008, *Slate* magazine applauded him as "an advocate for reducing dependency on fossil fuels." The Association for the Study of Peak Oil (ASPO-USA) created the Roscoe G. Bartlett "Speak Truth to Power" Award in his honor in 2008. It had previously awarded him the M. King Hubbert Award in 2006 for his leadership in the Congress to promote efficiency and conservation and alternative renewable sources of domestic energy to enable the United States to overcome the challenges to national security and economic prosperity of global peak oil. Congressman Bartlett is the cofounder and cochairman of the Congressional Peak Oil Caucus. He is also the cochairman of the House Renewable Energy and Energy Efficiency Caucus and Defense Energy Security Caucus. He is also a member of the Oil and National Security Caucus.

Beck, Dr. Chris
Email: chris.beck@eiscouncil.org

Chris Beck is the President of the Electric Infrastructure Security (EIS) Council. Dr. Beck is a technical and policy expert in several homeland security and national defense-related areas including critical infrastructure protection, cybersecurity, science and technology development, WMD prevention and protection, and emerging threat identification and mitigation.

Dr. Beck served as the Subcommittee Staff Director for Cybersecurity, Infrastructure Protection, and Science and Technology and was the Senior Advisor for Science and Technology for the House Committee on Homeland Security (CHS), where he worked from May 2005 to May 2011. Prior to CHS, he worked in the office of Congresswoman Loretta Sanchez for three years, beginning as a Congressional Science Fellow and then as a legislative assistant.

Before government service, Dr. Beck was a postdoctoral fellow and adjunct professor at Northeastern University. He earned a PhD in physics from Tufts University (2001) and a BS in physics from Montana State University (1994). He served in the Marine Corps Reserve for five years (1987–1992).

Birnbach, Mr. Curtis
Email: hudres@optonline.net

Curtis Birnbach is President, Chief Technology Officer, Board of Directors of Advanced Fusion Systems. Mr. Birnbach is an expert in electromagnetics, electron tubes, pulse power systems, materials science, manufacturing methods, intellectual property development, and a registered export agent. He holds 18 U.S. Patents, 8 foreign patents, and numerous patents pending. He has published papers with SPIE, IEEE, NASA, and EPRI. He has consulted and worked collaboratively with the Intelligence community, U.S. Army, U.S. Air Force, and other industrial and nongovernmental organizations.

Boland, Hon. Andrea
Email: sixwings@metrocast.net

State Representative Andrea Boland recently completed eight years (or four terms) in the Maine legislature. She is considered a leader in safety issues of electromagnetic radiation, especially from cellphones and smart meters. She became involved in electric grid protection against electromagnetic pulse and geomagnetic solar storms (GMD) at the suggestion of her regular scientific advisor. Her work is supported by several national experts. She has a BA degree from Elmira College and an MBA from Northeastern University, and studied at the Sorbonne and Institute of Political Studies in Paris. She was awarded the 2011 Health Freedom Hero Award by the National Health Federation for her work on health freedom and safety. Her legislative work has led to confronting major corporate interests on matters of transparency and regulatory capture, and public protections.

Boston, Dr. Wallace E.
Twitter: @wallyboston

Dr. Wallace E. Boston was appointed President and Chief Executive Officer of American Public University

System (APUS) and its parent company, American Public Education, Inc. in July 2004. He joined APUS as its Executive Vice President and Chief Financial Officer in 2002.

Dr. Boston guided APUS through its successful initial accreditation with the Higher Learning Commission of the North Central Association in 2006 and 10-year reaccreditation in 2011. In November 2007, Dr. Boston led the parent company of APUS, American Public Education, Inc. (APEI), to an initial public offering on the NASDAQ Exchange. During his tenure, APUS has grown to over 100,000 students majoring in more than 90 degree programs. Dr. Boston has authored and co-authored papers on the topic of online post-secondary student retention.

In addition to his service to the University, Dr. Boston serves as a Board Member of the Education Alliance, a nonprofit organization promoting public/private partnerships serving K-12 public schools in West Virginia. In his career prior to APEI and APUS, Dr. Boston served as either CFO, COO, or CEO of Meridian Healthcare, Manor Healthcare, Neighborcare Pharmacies, and Sun Healthcare Group.

Dr. Boston is a Certified Public Accountant and Certified Management Accountant. He earned an AB degree in History from Duke University, an MBA in Marketing and Accounting from Tulane University's Freeman School of Business Administration, and a Doctorate in Higher Education Management from the University of Pennsylvania's Graduate School of Education. In 2008, the Board of Trustees of APUS awarded him a Doctorate in Business Administration, honoris causa.

Briggs, Mr. Kevin
Email: kevin.briggs@hq.dhs.gov

Mr. Briggs is the DHS Team Chief, NCCIC (National Cyber Security and Communications Integration Center) and in that capacity leads various strategic and tactical activities regarding emergency communications including the SHARES program and provided technical guidance for the EMP protection of federal civilian communications systems.

Caruso, Mr. Michael
Email: carusomi54@gmail.com

Former sales executive at ETS-Lindgren, Mr. Caruso is now an independent consultant. Mr. Caruso is a globally recognized authority with 32 years of experience in the design and integration of EMC, EMP, TEMPEST, and SCIF facilities. Expert experience in managed accounts, business development, marketing strategy, sales planning, international business, start-ups, product launch, budgeting, proposal preparation, and project management. Extensive experience in the EMP protection of critical infrastructure, military, and government facilities. Successful track record in finding and developing strategic partnerships that create value and generate long-term revenue results. Consultation Service for (EMC, EMP, TEMPEST, and SCIF Facilities):

• advise architects, engineers, and end users in the design and integration;
• create risk analysis, basis of design, and specification documents;
• review designs, plans, and submittals;
• retrofit evaluations;
• assist in selecting materials and suppliers;
• prepare and present executive briefs and workshops;

- assist in preparing budgets;
- on-site quality inspections;
- member of InfraGard EMP SIG.

Chrosniak, BG (NYG-Ret) Kenneth
Email: kenneth.d.chrosniak.civ@mail.mil

BG (NYG-Ret) Ken Chrosniak enlisted in the Army in 1965, was later commissioned through Regular Army Officer Candidate School, and went on to complete 37 years of combined service in the Regular Army, Army Reserve, and National Guard, and was advanced to the rank of Brigadier General (New York Guard-Ret) by Governor George Pataki. He has an undergraduate degree in Secondary Education from Daemen College, and a Master's in Education from Saint Bonaventure University, and is an Army War College graduate.

Ken retired from active duty in 1998 while an instructor at the Army War College at Carlisle Barracks, where he then remained as a faculty instructor. In 2002 he was recalled to active duty service and served for two years on the Joint Chiefs of Staff, where he helped formulate the National Military Strategic Plan for the War on Terror. He then went on to serve in Iraq first at the U.S. Embassy, then later as Commander of the Abu Ghraib Forward Operating Base and detention facility. Upon release from active duty in 2005, Ken returned to Army War College instructor duties where, after one year, he was again recalled to active service for an additional two years as Chief of Staff of the Army Asymmetric Warfare Office located in the Pentagon. He presently is an Instructor in the Army War College Mission Command and Cyberspace Division.

He has served in varied command and staff assignments in the United States and overseas, including Vietnam, Bosnia, Kuwait, and Iraq. His most significant decorations are the Defense Superior Service Medal, the Legion of Merit, the Bronze Star Medal, the Good Conduct Medal, and the Combat Action Badge. Ken is an active Firefighter with Carlisle Fire and Rescue Company, and a member of the Carlisle Ambulance Company. However, he considers his present position within EMPact America and the EMP SIG as his most important calling in service to the Nation. Ken and his wife Gayle call Carlisle their home. He has two sons, Joshua and Christian, and a daughter Stephanie.

Cikotas, Mr. Bronius
Email:cikotasb@gmail.com

Mr. Cikotas is a senior research scientist and one of the nation's recognized experts in Electromagnetic Pulse technologies, with broad experience across the Department of Defense, the Intelligence Community, Inter-Agency, NATO, and Multi-National groups. He is responsible for the Executive level EMP-hardened communication system in Washington DC, the EMP hardening design of the Strategic Air Command and was the EMP Division Chief for the Defense Nuclear Agency. His work includes assessment of EMP national infrastructure vulnerability and EMP attack impact for both high altitude and ground burst EMP, as well as facility hardening, Command and Control system survivability, and tactical and strategic weapon system hardening.

Bronius Cikotas is a strategic advisor to the U.S. government on global EMP policies and on EMP threat assessment and delivery methods, attack scenarios and risk identification, infrastructure hardening and continuity planning, and has been a key contributor to EMP risk and risk-mitigation briefings to the President and Vice President of the United States, and to multiple U.S. Congressional Commissions and panels.

Cooper, Ambassador Henry F.
Email: hcooper@ara.com

Ambassador Henry F. Cooper is Chairman of High Frontier, Chairman Emeritus of Applied Research Associates, and a Director on the Boards of The Foundation for Resilient Societies, the London Center for Policy Research, and the EMP Task Force for Homeland and National Security. He previously served as Senior Associate of the National Institute for Public Policy and Visiting Fellow at the Heritage Foundation. At High Frontier, he is working with local, state, and federal authorities to provide effective defenses against ballistic missiles, particularly those that pose an existential EMP threat to all Americans. Since 1979, he has been appointed by the President to serve as Deputy Assistant Secretary of the Air Force with oversight responsibility for Air Force strategic and space systems (1979-1981); Assistant Director of the U.S. Arms Control and Disarmament Agency, backstopping all bilateral negotiations with the Soviet Union (1983-1985); Ambassador and Chief U.S. Negotiator at the Geneva Defense and Space Talks with the USSR (1985-1990); and Director of the Strategic Defense Initiative (SDI, 1990-1993). He served on numerous technical working groups and high-level advisory boards—including the Defense Science Board, the Air Force Scientific Advisory Board, U.S. Strategic Command's Strategic Advisory Group, the Defense Nuclear Agency's Scientific Advisory Group on Effects, and the Congressional Commission to Assess the U.S. Government's Organization and Programs to Combat the Proliferation of Weapons of Mass Destruction. He received the Defense Department's Distinguished Public Service Medal, the Defense Special Weapons Agency Lifetime Achievement Award, the U.S. Missile Defense Agency's Ronald Reagan Award, the U.S. Navy Aegis BMD Pathfinder Award, and Clemson University's Distinguished Service Medal. Ambassador Cooper taught at Clemson University and worked at Bell Telephone Labs, the Air Force Weapons Lab, R&D Associates, and JAYCOR. He earned BS and MS degrees from Clemson University and a PhD from New York University, all in Mechanical Engineering.

Donahue, Dr. Donald Jr., DHEd, MBA, FACHE, Lieutenant Colonel (Ret)
Email: donald.donahuejr@verizon.net

Dr. Donald Donahue is as an Advisory Member to the American Academy of Disaster Medicine (AADM) which is a unique volunteer organization dedicated to promoting and educating physicians and healthcare professionals concerning the critical importance of disaster medicine preparedness.

Dr. Donahue is a founding partner in Diogenec Group, a healthcare consultancy. He is a former Vice President for Department of Defense Programs with Comprehensive Health Services, a national workforce health management firm. Prior to that, he served as a Vice President with Jefferson Consulting Group in Washington DC.

Dr. Donahue served as the Deputy Surgeon for Policy and Fiscal Administration and the Medical Operations Officer for the Chief, Army Reserve. In this capacity, he was responsible for strategic planning, program development, and funding for medical operations within the Army Reserve and was the principal advisor for medical/dental readiness, policy, medical aspects of Homeland Security, and Bioterrorism issues.

Dr. Donahue was a principal planner for the Department of Defense (DoD) anthrax and smallpox immunization programs. He authored the first Joint Forces Command model for medical response to domestic weapons of mass destruction incidents and natural disasters and created the Joint/Interagency Civil Support Training Center (JISCTC) at Fort Dix, NJ, designed to provide training to DoD, VA, and other Federal departments in medical aspects of homeland security and weapons of mass destruction.

Dr. Donahue conceptualized and launched implementation of the organizational redesign of Army Reserve medical structure and creation of the Army Reserve Medical Command (AR-MEDCOM). He has served as Army Reserve principle for implementation of Health Insurance Portability and Accountability Act (HIPAA) and established Army Reserve Physician Assistant Training Program.

Certified in healthcare administration and a fellow of the American College of Healthcare Executives, and the Center for National Preparedness of the University of Pittsburgh, Dr. Donahue earned a BS in Sociology and Political Science from the University of the State of New York, an MBA from Baruch College, and a Doctorate in Health Education (DHEd) from A.T. Still University.

Donat, Dr. Terry L., MD, FACS, FICS
Email: terry.donat@outlook.com

Terry Donat, MD, is a dual-board certified Facial Plastic and Reconstructive Surgeon/OHNS and Medical Investigator practicing in Northern Illinois and greater metropolitan Chicago for the past 15 years. Terry trained as a Biochemist and received his Medical Degree in Philadelphia in 1991. Terry is a written-exam reviewer and oral board examiner of U.S. and Canadian surgeons for the American Board of Facial Plastic and Reconstructive Surgery. He has extensive experience in facial reconstruction and managing blunt/penetrating head and neck trauma. Terry is the first physician certified as an Illinois Professional Emergency Manager. He is trained in acute Radiation Emergency Medicine, the Medical Management of Chemical and Biological Casualties and as a past National Disaster Life Support Instructor. He is appointed IEMA RACES Regional Radio Officer on the IEMA State Team—Region 2.

Terry is the first physician to complete the graduate program in Veterinary Homeland Security from Purdue University's National Biosecurity Resource Center. He is a lifetime member of the Special Operations Medial Association (SOMA) and is currently focused on novel means for mitigating heat stress in austere environments and extending force protection while wearing PPE. Terry serves as the Acting Sector Chief—Health and Public Health for InfraGard Chicago; serves aside James Terbush MD MPH as co-chair of the Healthcare Industry Advisory Group of the InfraGard Electromagnetic Pulse (EMP) Special Interest Group; and serves as a member of the Healthcare and Public Health Sector Coordinating Council—DHS/DHHS. He is keenly interested in *Biosecurity and Medical Intelligence* and assessing the threats, risks and vulnerabilities of evolving dual-use technologies.

Franks, Congressman Trent
Email: https://franks.house.gov/contact-me/email-me

Congressman Franks is serving his fourth term in the U.S. Congress, representing the Second District of Arizona. He serves on the Armed Services Committee and the Judiciary Committee, where he is the ranking member of the Constitution Subcommittee. He is an active member of the Republican Study Committee and is an Executive Committee Member of the Tom Lantos Congressional Human Rights Commission, and co-founder and co-chairman of the International Religious Freedom Caucus.

Congressman Franks also serves on a host of task force and caucus groups, including the House Working Group on Judicial Accountability, the Education Freedom Caucus, the House Working Group on Waste, Fraud and Abuse, the Liberty Caucus, the Human Rights Caucus, the India Caucus, the Anti-Terrorism Caucus, and is co-founder of the Israel Allies Caucus (IAC) and co-chair of the EMP Caucus.

Glynn, Captain Arthur
Email: Art_JFK_Glynn@yahoo.com

Captain Art "JFK" Glynn is the Navy Emergency Preparedness Liaison Officer (NEPLO) to U.S. Northern Command (USNORTHCOM) and a Deputy Command Center Director in the North American Aerospace Defense Command (NORAD) and USNORTHCOM Command Center (N2C2).

A native of Portland, Oregon, he graduated from Oregon State University in 1985 with a Bachelor of Science in Finance and International Business. He reported to Aviation Officer Candidate School in December 1985 and was commissioned an Ensign in April 1986. He was designated a Naval Aviator in August 1988.

CAPT Glynn's first assignment was to Helicopter Combat Support Squadron 1, flying the CH-53E Super Sea Stallion. During this tour he deployed to the Persian Gulf for Operations Desert Shield and Desert Storm. Upon his return, he was assigned to a variety of increasingly responsible positions, culminating as the COMNAVAIRPAC Force Carrier Operations Officer. It was here that he also and earned a Master's of Science degree in Business Administration, International Business. Subsequent to his release from active duty in April 1995, CAPT Glynn joined the Navy Reserve and served in numerous leadership roles including as Commanding Officer of NR NAVAIRWINGSPAC 1094 and NR COMTHIRDFLT Headquarters units. While serving at COMTHIRDFLT, CAPT Glynn was recalled to active duty supporting COMTHIRDFLT's Joint Maritime Operations Center (JMOC), where he was selected as JMOC Director. Next, he joined the emergency preparedness mission and was selected as the NEPLO for FEMA Region VIII, adjunct Fellow to the USNORTHCOM Chair at the National Defense University and now at NORAD and USNORTHCOM.

In his civilian career, CAPT Glynn has been a President/CEO, entrepreneur and consultant. He now consults for Booz, Allen Hamilton in Colorado Springs, Co. in support of the NORAD and USNORTHCOM Interagency and S&T Directorates.

Hunt, Mr. David
Email: Davehunt001@gmail.com

Mr. Hunt has over 28 years in emergency response and national level disaster preparedness and planning, working with all 56 states and territories. Over the past several years, Mr. Hunt has conducted dozens of seminars focused on logistics planning, functional/special needs, emergency public information and Joint Information Center operations, and strategic planning studies. Mr. Hunt managed Improvised Nuclear Device (IND) planning for FEMA's National Planning Coordination and Assistance Branch, including the development of a pilot public education campaign for preincident IND messaging. He has led a team of 10 employees and dozens of consultants and subcontractors, including scientists and other experts from Lawrence Livermore National Labs and Virginia Tech's Center for Technology Security and Policy, in developing IND evacuation strategies and software focused on nuclear planning factors for individual communities. He has worked with the New York City region and many other jurisdictions in the development of response plans, exercises and several pilot projects.

Since 2010, he has conducted energy incident planning and grid collapse seminars, including for the Congressional EMP Caucus, National Defense University, Maryland EMA, Johns Hopkins, Energy Infrastructure Security (EIS) Council, and the FBI InfraGard EMP Special Interest Group. He serves on the Electric Grid Protection Executive Steering Committee of the EIS Council. Mr. Hunt earned a BS from Virginia Tech and has over 4,000 hours of specialized training in all aspects of emergency response, incident management, and disaster planning.

Johnson, Senator Ron
Email: https://www.ronjohnson.senate.gov/public/index.cfm/email-the-senator

Senator Johnson came to Washington because the federal government is bankrupting America. He thinks it is important for citizen legislators to ally with those who are seriously facing that reality. Ron's manufacturing background has taught him to attack the root cause of a problem, not mere symptoms. He believes huge deficits, slow economic activity, high unemployment, and woefully inadequate job creation are severe symptoms of the problem—but not the root cause. The ever expanding size, scope, and cost of government is.

Ron is Chairman of the Homeland Security and Governmental Affairs Committee, and also serves on the Budget, Foreign Relations, and Commerce, Science and Transportation committees. He resides in Oshkosh, Wisconsin with his wife Jane. They have three children. He cofounded—PACUR—producing plastic sheet for packaging and printing applications. He attended the University of MN and has a BSB in Accounting, and an MBA.

Kishi, Mr. Arnold
Email: akishi@hawaii.edu

Arnold Kishi is a management consultant and adjunct faculty supporting University of Hawaii and National Disaster Preparedness Training Center programs. He earlier retired after serving 30 years with the State of Hawaii and the East-West Center. Arnold has been on several national commissions and councils, and currently serves with the Harvard Policy Group for Network-Enabled Services and Government; and on Boards of the Center for Internet Security, CIO Council of Hawaii, University of Hawaii Alumni Association, and FBI's Honolulu InfraGard. Arnold has MBA and BA degrees from University of Hawaii, and completed post-graduate Executive Programs at the National Emergency Training Center (EMI), Stanford University, and Harvard University's Kennedy School of Government.

Krieg, Dr. Richard, PhD
Email: rkrieg@thehorizonfoundation.org

Dr. Krieg has specialized in health system management, state and regional policy development and public policy development and implementation. Over the past 14 years, he has focused on the improvement of community resilience and the promotion of joint public/private sector emergency response planning. He is founding Chairman of the Community Emergency Response Network. He is immediate past President and a former Trustee of The Horizon Foundation. The Foundation is the largest health philanthropy in the Mid-Atlantic states with a multi-faceted community health agenda. Over the years, the Foundation has provided multiple grants designed to improve communications inter-operability, first responder training, tabletop, functional and full bore emergency exercises, and the improvement of hospital and health department readiness.

Dr. Krieg is former Health Commissioner for the City of Chicago. The Chicago Department of Health has a staff of approximately 2,000 people and an $85 million annual budget. In that role, he focused on Level-3 trauma center improvement and public health/hospital coordination in emergency response. He is former Executive Director of the Chicago Institute for Metropolitan Affairs, an urban policy and community action facility that spearheaded health system reform, reduction in hospital emergency department overcrowding, Cook County Hospital reform and other issues. For four years, he held the position of Director of Policy Analysis and Planning for the Metropolitan Chicago Healthcare Council.

He received his doctorate from the University of Chicago where he specialized in health policy and administration. He earned a master's degree from the University of Chicago's Harris Graduate School of Policy Studies. He is a graduate of the Executive Management Program at the Harvard Business School. Among other commendations, he is recipient of the Illinois Department of Public Health's "Award of Merit" and the Illinois Public Health Association's "Presidential Award."

Lasky, Ms. Mary
Email: mary.lasky@jhuapl.edu

Ms. Lasky volunteers as the Chairman of the Howard County Community Emergency Response Network (CERN) and currently is Program Manager for Business Continuity Planning for the Johns Hopkins University Applied Physics Laboratory (JHU/APL). She leads APL business continuity planning including their pandemic influenza response effort. She also coordinates the APL Incident Management Team, which is responsible for managing any disaster that disrupts normal operations for more than 48 hours. CERN's mission is to spearhead the development of a community-wide disaster response plan to ensure maximum preparedness in the event of a terrorist attack or major natural emergency. The unique community readiness program was initiated after the events of 9/11 as a partnership between the Horizon Foundation, Howard County government, and key community agencies in Howard County. The effort supports government disaster planning through coordination of the emergency plans and resources of participating members. CERN functions include planning, a high level of inter-agency coordination, the development of tabletop exercises, disaster plan review, shelter planning, and communications enhancement. In addition to her APL roles, Lasky is the chair of the Pandemic Influenza Education and Communication Group for the Johns Hopkins Office of Critical Event Preparedness and Response (CEPAR), which is responsible for pandemic planning for all Hopkins Institutions. She is also on the adjunct faculty of the Johns Hopkins University, Whiting School of Engineering, teaching in the Technical Management graduate degree program.

Leighton, Mr. Cedric
Email: cedricleighton@mac.com

Cedric honed his analytical and leadership skills during a 26-year career as an intelligence officer in the U.S. Air Force. He witnessed the fall of the Berlin Wall, oversaw critical Special Operations missions, established key partnerships with nations in South and Southeast Asia and deployed five times to the Middle East. He served at every command echelon from small-deployed elements to the Joint Staff at the Pentagon, where he was the Deputy Director for Warfighter Support and Integration in the Intelligence Directorate. His last military assignment was as the National Security Agency's Deputy Director for Training. His numerous military awards include the Defense Superior Service Medal, the Bronze Star (for his work in Operation Iraqi Freedom), the Defense Meritorious Service Medal, and seven Meritorious Service Medals. He retired as a Colonel in 2010.

Since founding Cedric Leighton Associates, Cedric has become an internationally known strategic risk expert. He has appeared on Fox Business Network, Fox News Channel, Bloomberg TV, CNN, C-Span, MSNBC, BBC, CCTV (Chinese Central TV), CTV (Canada), ABC (Australia), RT America, Fox 5 DC, Fox 29 Philadelphia, as well as Washington DC's News Channel 8, where he is a regular contributor to the "Capital Insider" program. He has written articles on cyber strategy, national security and management topics for AOL Defense, Business Excellence Magazine, MWorld, Leadership Excellence Magazine, theStreet.com, CNBC.com, JenningsWire, and the Huffington Post. On radio, he has appeared on national shows such as the Jim Bohannon Show and the Leslie Marshall show, as well as numerous regional radio talk shows. Internationally, he has been a guest on the Voice of America and was quoted by the French national newspaper Le Figaro. He was a panelist for three Bloomberg Cyber Security Defense and National Security conferences. He is a National Journal "National Security Insider," one of 89 nationally recognized experts who contribute insights every week.

Cedric graduated magna cum laude from Cornell University and earned a Master's Degree in International Studies from Angelo State University. He speaks German, French and Spanish and has traveled extensively in Europe, the Middle East, Asia, and Latin America.

MacLellan, Mr. Thomas
Email: tmaclellan@nga.org

Thomas MacLellan is Director of the National Governors Association Center for Best Practices Homeland Security and Public Safety Division, where he leads the Center's work on homeland security and public safety policy issues. MacLellan led the creation of the NGA Resource Center for State Cybersecurity, the first national effort designed to focus on state cybersecurity policy issues and the role of governors. He has authored numerous publications on issues related to cybersecurity, prescription drug abuse, mass evacuation, prisoner reentry, forensic DNA, juvenile justice, school violence, and other related issues. He earned a BA from the College of the Holy Cross and is a graduate of the Naval Postgraduate Executive Leaders Program.

Manto, Mr. Charles
Email: cmanto@stop-EMP.com

Mr. Manto is CEO of Instant Access Networks, LLC a consulting and R&D firm that produced independently tested solutions for EMP protected micro-grids. He received six patents in telecommunications and computer mass storage and EMP protection and assisted other entrepreneurs and investors with their intellectual property strategies. Developed valuation methodology accepted by the U.S. DoD, countries, and companies participating in industrial defense conversion. Facilitated due diligence of over 200 deals, managed a venture capital service, a revolving loan fund, an economic development corporation, a computer mass storage manufacturer, and broadband CLEC. Mr. Manto has also founded and leads InfraGard National's EMP SIG. He received his BA and MA from the University of IL at Urbana/Champaign.

Moeller, Commissioner Philip D.
Twitter: @PMoellerFERC

Commissioner Philip D. Moeller is serving his second term on the Commission, having been nominated by President Obama and sworn in on July 16, 2010, by Congresswoman Cathy McMorris Rodgers (R-Washington), for a term expiring June 30, 2015. He was first nominated to FERC by President George W. Bush in 2006 and sworn into office on July 24, 2006, by Chief Justice of the United States John Roberts.

From 1997 through 2000, Mr. Moeller served as an energy policy advisor to U.S. Senator Slade Gorton (R-Washington) where he worked on electricity policy, electric system reliability, hydropower, energy efficiency, nuclear waste, energy and water appropriations, and other energy legislation.

Prior to joining Senator Gorton's staff, he served as the Staff Coordinator for the Washington State Senate

Committee on Energy, Utilities, and Telecommunications, where he was responsible for a wide range of policy areas that included energy, telecommunications, conservation, water, and nuclear waste.

Before becoming a Commissioner, Mr. Moeller headed the Washington DC, office of Alliant Energy Corporation. Prior to Alliant Energy, Mr. Moeller worked in the Washington office of Calpine Corporation.

Mr. Moeller was born in Chicago, and grew up on a ranch near Spokane, Washington.

He received a BA in Political Science from Stanford University.

Newman, Major General Robert Newman (USAF, Ret.)
Email: robertnewmanjr1@gmail.com

Major General Robert B. Newman, Jr. served as Adjutant General of Virginia from January 2006 to May 2010, where he was responsible for the combat readiness of units, and the administration and training of more than 8,200 Virginia Army and Air National Guard personnel. Prior to his appointment as the Adjutant General, he served as the Deputy Assistant Homeland Security Advisor to Virginia Governor Mark Warner where he oversaw Virginia's critical infrastructure protection program and the Commonwealth Preparedness Working Group. He also served as Vice Director for Operations, Plans, Logistics and Engineering (J3/4V) at U.S. Joint Forces Command; and Commandant of the National Guard's first joint unit combining both Air and Army Guardsmen from Virginia and North Carolina at U.S. Joint Forces Command. After the attacks of September 11, he was called to active duty serving as the liaison officer for the National Guard to the U.S. Joint Forces Command as Chief of the Special Programs Branch, J3/Domestic Operations Division, National Guard Bureau.

Nordling, Mr. Gale K.
Email: gnordling@emprimus.com

Mr. Nordling, President and CEO of Emprimus, has 35 years experience as an engineer, risk manager, risk management and insurance consultant, and expert witness. He has been involved with the preparation, negotiation, settlement, litigation, arbitration, mediation, and insurance coverage of over $500 million of claims and contract disputes for engineers, contractors, suppliers and owners including universities, hospitals, states, airlines, casinos, and utilities.

Mr. Nordling has been employed by a nuclear utility, a disaster recovery company, national construction company, and international risk management firm. Mr. Nordling served on a national committee to create a national pooled inventory and management of safety-related spare equipment for all nuclear plants. The disaster recovery company included some of the largest upper Midwest companies including ConAgra, Cargill, Northwest Airlines, National Car, Gelco, Minnegasco, Northern States Power, and various insurance and banking institutions.

Popik, Mr. Thomas (Tom)
Email: thomasp@resilientsocieties.org

Thomas Popik is chairman of the Foundation for Resilient Societies, a nonprofit group dedicated to the protection of critical infrastructure against infrequently occurring natural and manmade disasters. He is principal author of a Petition for Rulemaking submitted to the Nuclear Regulatory Commission that would require backup power sources for spent fuel pools at nuclear power plants. Previously, as a U.S. Air Force officer, Mr. Popik investigated unattended power systems for remote military installations. Mr. Popik graduated from MIT with a BS in mechanical engineering and from Harvard Business School with an MBA.

Reeves, Senator Bryce Reeves
Email: district17@virginia.state.gov

State Senator of the 17th District of Virginia, Bryce Reeves is a graduate of Texas A&M University with a Bachelor of Science degree where he was recognized for his academic achievement with the coveted Distinguished Military Graduate designation. Upon graduation Bryce was commissioned a Second Lieutenant in the U.S. Army and served as an Airborne Ranger in the Infantry. Bryce recently completed the Senior Executives in State and Local Government program, July 2013, Harvard University, John F. Kennedy School of Government. In addition to his committee assignments, Bryce was chosen to serve as the Co-Chair of the Military Caucus. He is currently the Senate representative on the Virginia Military Advisory Council to the Governor (VMAC) and was recently appointed to the reconstituted Commission on Military Installations and Defense Activities (COMIDA) (for Sequestration and BRAC 2015). Bryce serves on the Rappahannock River Basin Commission, is a standing member of the National Conference of State Legislatures (NCSL)-Task Force on Military and Veteran's Affairs, and the NCSL Human Service and Welfare committee. He also serves on the Virginia Sesquicentennial of the American Civil War Commission, the Veterans Services Board of Virginia, the advisory Committee on Juvenile Justice, the Virginia Commission on Youth, and a member of the 2012/2013 Offender Population Forecast Policy Committee. Bryce was recently appointed to the Executive Board of the Crime Commission of Virginia and the Energy and Environment Committee for the Southern Legislative Conference of the Council of State Government.

Rich, Dr. Paul
Email: pauljrich@gmail.com

President, Policy Studies Organization, Washington DC. Adjunct Professor, George Mason University. Visiting Fellow, Hoover Institution, Stanford University.

As President of the Policy Studies Organization, Dr. Rich is responsible for appointing the editors of 12 journals and of book series published for the PSO by Wiley-Blackwell, Berkeley Electronic Press, Rowland & Littlefield, and Global Information Company. The journals and their editorial offices include Policy and Internet (University of Oxford), Review of Policy Research (Ryerson University, Toronto), Digest of Middle East Studies (University of Wisconsin), Risk, Hazards & Crisis in Public Policy (Pacific Disaster Center, University of Hawaii), Policy Studies Journal (University of Colorado, Denver), Politics & Policy (Universidad IberoAmericana, Mexico City), Latin American Policy (Tecnologico de Monterrey, Mexico City), Asian Politics & Policy (Asian Center, University of the Philippines), World Medical and Health Policy (George Mason University), Poverty & Public Policy (University of Missouri, Kansas City), Proceedings of the PSO (PSO Headquarters, Washington DC).

The PSO publishes *The Yearbook of Policy Studies* (University of Oklahoma), book series on *Public Policy, The Middle East and China* (University of San Francisco), and hosts the Dupont Summit on Science & Technology Policy—an annual December conference in Washington DC at the Carnegie Institution, the Middle East Dialogue at the Washington Club each February, practitioner workshops at annual meetings of APSA, Westminster style parliamentary debates in cooperation with Tulane University, and as a sponsor with Oxford

University's Internet Institute of biannual Internet policy world conference.

The Policy Studies Organization includes more than 3,600 universities and institutions as well as individuals in more than 93 countries. From its headquarters in an historic house near Dupont Circle in Washington, home of the families of President James Garfield and the labor leader Samuel Gompers, the society organizes conferences, seminars, the publication of journals and books, and an increasing Internet presence. The PSO offices have a notable antiquarian collection relating to its history and the Dupont Circle area. Interns are placed with PSO by Cornell, Penn State, and the Washington Center; PSO visiting professors are provided office space.

Dr. Rich is a Kinsmen Scholar, Tonbridge School, England. Jr. Common Room Chr., Dunster House, Harvard College. AB cum laude, EDM, Harvard. PhD, University of Western Australia. Harvard Mountaineering Club (life). Secretary Harvard Liberal Union.

Riggi, Mr. John
Email: John.Riggi@ic.fbi.gov

FBI Section Chief Riggi is a highly decorated 27-year veteran of the FBI, and currently serves as the Section Chief for the Cyber Division Outreach Section where he leads the development of mission critical partnerships with the private sector. Previously, Mr. Riggi served as an Assistant Special Agent in Charge for the Washington Field Office's, Intelligence Division. In 2013 Mr. Riggi was selected to lead the development of the FBI's Cyber Financial Pursuit Team. Previously, Mr. Riggi served for four years as the National Operations Manager for the FBI's Terrorist Financing Operations Section and two years at CIA's Counterterrorism Center (CTC).

Prior, Mr. Riggi served 16 years in the FBI New York Field Office as a Case Agent, Supervisor of the High Intensity Financial Crimes Area (HIFCA) Task Force and Supervisor of the Terrorist Financing Squad. In New York, he developed and led the FBI's first undercover operation into Russian Organized Crime, was the first to use sting money laundering transactions in terrorism financing cases and initiated the Alavi Foundation terrorism financing case which resulted in the largest counterterrorism seizure of assets in U.S. history. Mr. Riggi also acted in an undercover capacity posing as an organized crime money launderer to penetrate and expose the multi-billion dollar commercial check cashing industry in New York City and its connection to organized crime money laundering and bank corruption. Mr. Riggi also served as an operator on the FBI NY SWAT team for eight years. Mr. Riggi began his career as an FBI Special Agent in the Birmingham Office in 1988.

Mr. Riggi is a recipient of the FBI Director's Award for leading a highly successful classified terrorism financing interdiction program which was responsible for preventing multiple terrorist attacks in a foreign ally's nation. Mr. Riggi is also a recipient of the CIA George H.W. Bush Award for Excellence in Counterterrorism, which is the CIA's highest counterterrorism award, for greatly expanding FBI/CIA joint counterterrorism operations and cooperation.

Rutledge, Mr. Robert
Email: robert.rutledge@noaa.gov

Mr. Rutledge is the Lead of NOAA Space Weather Forecast Office. The role of the Space Weather Prediction Center is to deliver space weather products and services that meet the evolving needs of the nation. The Space Weather Prediction Center gathers, in real time, the available data that describes the state of the Sun,

Heliosphere, Magnetosphere, and Ionosphere to form a picture of the environment from the Sun to the Earth. With this information, forecasts, watches, warnings and alerts are prepared by the Space Weather Prediction Center and issued to anyone affected by space weather.

Sheppard, Dr. Ben, PhD
Email: ben@sheppard.net

Dr. Sheppard's research focuses on the political and psychological consequences of manmade and natural disasters on population centers, and risk communication strategies to elicit desired behaviors. This includes investigating societal ripple effects from terrorist attacks and risk communication strategies. He is a Professorial Lecturer at the Elliott School, George Washington University, and Affiliate Faculty Member at the Center for Health and Risk Communication, University of Maryland.

He was Principal Investigator for an Improvised Nuclear Device Communications study, and a Co-Investigator on a three-year war game simulation project at King's College London, and researcher on developing risk communication guides for emergency managers as part of a risk communication training and simulation project at START. As a futurist at the Institute for Alternative Futures, Dr. Sheppard employs scenarios to identify national security challenges and opportunities. Recent projects include cyber security scenarios for robots looking out to 2030. Dr. Sheppard's research has been funded by the Department of Homeland Security, Homeland Security Studies and Analysis Institute, the UK government, and pharmaceutical companies. He has also collaborated with CREATE at the University of California.

He is the author of Psychology of Strategic Terrorism (Routledge, 2009), based on his PhD he received at King's College London. He combines international relations and risk analysis techniques to enhance homeland security policy development and decision making.

Dr. Sheppard also designs and runs war game simulations for the public and private sector to examine new homeland security initiatives, business strategies, and innovative decision-making systems. As a defense analyst at Jane's Defense, he managed and commissioned defense Special Reports titles, was the Director of the Jane's Annual Ballistic Missile Proliferation Conference series, and Editor of the Sentinel threat assessment series.

Sheppard is also an Adjunct Fellow of the Potomac Institute for Policy Studies. Sheppard received his Masters in Strategic Studies and BSc Econ (Hons.) in International Politics at Aberystwyth University in Wales. Sheppard's research is published in a number of publications including Jane's Defense Weekly, and he has authored chapters for various books on missile proliferation and terrorism.

Stockton, Dr. Paul N.
Email: pstockton@cloudpeak.sonecon.com
Assistant Secretary of Defense for Homeland Defense and Americas' Security Affairs

Paul N. Stockton was nominated by President Barack Obama to be the Assistant Secretary of Defense for Homeland Defense and Americas' Security Affairs on April 28, 2009, and was confirmed by the Senate on May 18, 2009. In this position, he is responsible for supervising the Department of Defense's homeland defense activities (including Defense Critical Infrastructure Protection and other mission assurance efforts), defense support of civil authorities, domestic crisis management, and Western Hemisphere security matters.

Assistant Secretary Stockton received a bachelor's degree from Dartmouth College Summa Cum Laude in 1976, and a doctorate in government from Harvard in 1986. From 1986 to 1989, Assistant Secretary Stockton served as legislative assistant to Senator Daniel Patrick Moynihan, advising the senator on defense, intelligence, and counter-narcotics policy, and serving as the Senator's personal representative to the Senate Foreign Relations Committee. From 1989 to 1990, Assistant Secretary Stockton was a Postdoctoral Fellow at Stanford University's Center for International Security and Cooperation. During his graduate studies at Harvard, he served as a research associate at the International Institute for Strategic Studies in London.

Terbush, Dr. James W., MD, MPH
Email: jim.terbush@martin-blanck.com

James Terbush MD, MPH is a Senior Partner with Martin, Blanck & Associates and joined the firm in April 2014. From 2006 to 2009, Captain Terbush USN, served as Command Surgeon to North American Aerospace Defense Command (NORAD) and U.S. Northern Command (USNORTHCOM). In this role, he served as the Medical Advisor to the Commander and was responsible for the integration of Department of Defense medical assets internally and with other agencies in support of military response to civilian disasters combating terrorism and protecting Americans. From 2009 to 2011, Dr. Terbush served as the Fleet Surgeon for Commander, U.S. Naval Forces Southern Command. He was deployed forward to Port au Prince in response to the devastating earthquake disaster in Haiti, integrating DoD medical capabilities into the overall International response. Dr. Terbush's final assignment before retiring from military service was with the Science and Technology Directorate at NORAD and USNORTHCOM where he served as the lead for medical innovations.

In more than 30 years of Government service Dr. Terbush was the physician to U.S. personnel in more than 80 countries. He is published in scientific journals on; Influenza and Air Travel, Mass Fatalities Management and Public Health Consequences of a Cyber Attack. Dr. Terbush currently serves on multiple boards of directors; Public Health for El Paso County Colorado, Peak Military Care Network (veterans affairs), graduate medical education (University of Colorado), and disaster medicine. He is also an advisor to the National Academy of Sciences Institute of Medicine Forum on Disaster and Public health and is the Past President of the American Academy of Disaster Medicine.

Dr. Terbush received his MD degree from the University of Colorado and a Master's in Public Health from the University of California, Los Angeles.

Introduction and Opening of the 2014 InfraGard National EMP SIG Sessions of the Dupont Summit at the Whittemore House in Washington DC

Keynote Address by Senator Ron Johnson, with Introduction by Dr. Wallace E. Boston after Opening Remarks by Mr. Charles (Chuck) Manto

To see video recording, use this web address: https://youtu.be/4_Uhlco73Zg

Dupont Summit on Friday, December 5, 2014

Mr. Charles Manto introduces the InfraGard National EMP SIG and how it fosters resilience for local communities who will want to be better prepared to cope with high-impact disasters. He then introduces Dr. Wallace E. Boston, President and CEO of American Public University System. Dr. Boston introduces Senator Ron Johnson, who serves on the Committees on Budget, Commerce, Science and Transportation, Foreign Relations, Homeland Security and Government Affairs (now as Chairman), and Small Business and Entrepreneurship. Senator Johnson covers the implications of high-impact events such as a collapsed economy due to run-away debt and electromagnetic pulse (EMP). Senator Johnson then discusses the steps Congress needs to take in order to systematically address the EMP issue effectively.

CHARLES (CHUCK) MANTO: *(Note that this portion of the introduction is not on the accompanying video that begins after this segment.)* Good morning. My name is Chuck Manto. I head up an organization called InfraGard National's Electromagnetic Pulse Special Interest Group (EMP SIG). We sometimes call it the "every major problem special interest group." And the reason we do is that, as you know, InfraGard is a 501c3 organization of individuals, volunteers who are involved in critical infrastructure of any kind, who care about its development and security. And our group that formed just a few years ago has a special task. And that is, we look at anything that could impact critical infrastructure nationwide for more than a month. It could be a high-altitude nuke-burst EMP. It could be a space weather event. It could be a cyber attack. It could be a pandemic, a coordinated physical attack, and so on.

One of the reasons why it's really important that we do this is that this is a culturally and a qualitatively different kind of a problem. Normally in our country, when we experience something like Katrina, the rest of the country can come to our aid. And that's a wonderful thing. And we have been doing such a great job at improving how well we do that, we wind up having something perverse that happens. When the economy is hard, and life is tough, we wind up saying, "Why should I lift a finger to be more prepared or spend a nickel doing it, when these wonderful people are going to come rescue me anyway?" You know, the Feds, the state, even guys at the county level.

And so therefore, the things that are sometimes the most important for all of us to do, we don't bother to do right now. And unfortunately, it could be too late. So one of the benefits of doing what we're doing today is that, by looking at this class of problems that could make the entire country experience a Katrina simultaneously, it becomes obvious then that we all must do more to become more locally resilient in our community, so we not only can take care of ourselves, but our neighbors.

So then, when our heroes come to rescue us, which might even be a year, we're going to be just fine. And we're going to actually take care of them. In fact, Clara Barton—Remember, she was the woman who used to bail out the Union Army. She wasn't even a person with the right to vote. She didn't wait to get bailed out by the Feds. She bailed out the Feds. And that's what we're trying to do here.

55

(video recording link begins here:)

CHUCK MANTO: So we're very honored today to have Senator Johnson, who will give us some opening remarks in just a moment about this wonderful new Critical Infrastructure Protection Act (CIPA) and other things that he may want to care about, to talk to us about. But I want to take a moment to introduce the person who will introduce him.

As you know, we're here at the Dupont Summit. And the Dupont Summit is organized by all those organizations that are basically universities with public policy departments, the Policy Studies Organization. And one of the big sponsors of this group is the American Public University System. And Dr. Boston has been a key leader in this organization. He not only brought them through the process of accreditation, they now have well over 100,000 students. They're one of the premiere online organizations in the world.

We're working with their library folks to make certain that we can synchronize the "for official use only" library that InfraGard members who are cleared by the FBI can use, and all the Homeland Security kinds of things that they offer, along with the Naval Post-Graduate College. So we're very honored to have him here today to give the introduction to Senator Johnson. And I'd like to introduce to you someone we're very enthused to be connected with through the DuPont Summit, Dr. Wallace Boston.

[applause]

DR. WALLACE BOSTON: Thank you very much. I'm sure that we'd like to hear our speaker more than me, so I'll try to be brief with my biography. Senator Ron Johnson has been a hard worker all of his life. As a boy, he mowed lawns, shoveled snow, delivered papers, and even caddied for a few extra dollars. At the age of 15, he took his first job where he paid money into Social Security as a dishwasher in a Walgreens grill, later moving up to the job as night manager before he was 16.

He gained early acceptance into the University of Minnesota, skipping his senior year of high school, and worked full-time while obtaining a degree in business and accounting. You can see where this is going. After graduation, he worked as an accountant and earned his MBA at night. He's a highly successful entrepreneur and business leader. He brings a wealth of experience and private sector perspective to the Senate. He's fairly critical of the expanding size, scope, and cost of government regulation.

The Senator serves on the Committees on Budget, Commerce, Science and Transportation, Foreign Relations, Homeland Security and Government Affairs, Small Business and Entrepreneurship. He lives in Oshkosh, Wisconsin with his wife Jane. They have three children. We are immensely proud that he's taken time from all of his activities to open the conference. Please welcome Senator Ron Johnson.

[applause]

SENATOR RON JOHNSON: Well, thank you, Wally. Thank you, Chuck. I thank all of you for being involved. I think this is a relatively important issue here. This is also relatively intimidating crowd to be speaking in front of. I'll tell you what I'm not. I'm not an engineer. I'm not anywhere near smart as you guys. I mean you guys are the smartest people in the room here. I'm an accountant. I'm a business guy.

So what I want to bring to this discussion is how do you solve a problem? Kind of a businessman's perspective. Because, you know, I've been in accounting. But what I've primarily been is a manufacturer for 31 years. And being in manufacturing, you're always solving problems. And there is an organized approach to solving a problem. It generally starts, by the way, with admitting you have one—[laughter] which is kind of a big deal; and, then properly defining it.

And I'd like to use a couple examples in terms of government, in terms of why it's so difficult to solve some of these problems. One thing I do try and do when I'm talking to crowds, especially as a business person talking to business people, to try and get them to understand, in some way, shape or form, the dysfunction that is the federal government, is you know, convey the differences between the business world and government—but, also to convey the similarities.

And here is one thing that the private sector and the public sector have in common, they're motivated by the same factor. They both want to grow. The difference in the private sector is, in order to grow, you have to succeed at something. You have to actually produce a successful product, a successful service. That's how you grow. And in the public sector, let's face it, failure allows you to grow.

Now I make that point because I'll come back to it, because I think it's crucial, in terms of why we don't grapple with some of these problems and why we address others. So my first understanding of EMP really dates back a long time. I heard about this—you know, some nuclear blast, it could be catastrophic. Oh yeah. But what's the likelihood of that? I first really became aware of it a few months ago, when a gentleman named Bob Pfaltzgraff and Hank Cooper visited my office and presented me a little book, which I started to read. And I didn't really like reading it. And I got a little concerned.

And so that really began my journey here. But the reason I came to Washington DC was to address some other problems. And I want to talk about those other problems to put in context what it is we need to do to start addressing this problem. Now I ran for office because Washington passed ObamaCare, which is basically a government takeover healthcare system which I think is going to do a great deal of harm to a healthcare system, I think it's an assault on our freedom. I mean that was the Number One issue.

But, you know, then along the way, too, I was a little concerned about the fact that we're mortgaging our children's future. Now I'm an accountant. As I've gotten here, I've really tried to delve into the full extent of the problem. I just want to ask a question here. Anybody—Anybody in the room here know what our 30 year projected deficit is? Ever heard of it? There's a reason why you've never heard of it, because nobody wants to tell you what it is.

Congressional Budget Office (CBO) actually does project on these things. But they don't put it in dollar terms where you can actually understand it. Some people talk about, you know, the unfunded liabilities, you know Net Present Value back over 75—nobody can understand that. Nobody can relate to it. So in my dealings with the White House, I was one of the small group of about eight Republican Senators trying to work with the White House, trying to find some common ground, at least take the right steps in solving the problem.

I came up with the concept that we really ought to look and define the problem properly. Because let's face it, you know, talking about a problem-solving process, America has not collectively admitted we have this enormous debt and deficit issue. They tell you, "Yeah, it's unsustainable," but no one can really define it. And we haven't properly defined it. We don't have a 10-year budget window problem. We have a 30-year demographic problem. All the Baby Boom generation, we're retiring at the rate of, you know, 10,000 people per day. We've promised all these benefits, because that's an easy thing to do. We don't have any way to pay for it.

So anyway, CBO actually projects these things out. They don't do it in dollars. They do it as a percentage of gross domestic product (GDP). Nobody really understands that. So we converted those projections into dollars. Let me give you the bad news. Their baseline shows that—or projects our deficit over the next 30 years to be $66 trillion dollars. But the ultimate fiscal scenario is $127 trillion dollars.

Now, when you annualize that, when you analyze it, when you take a look at really what's been our 20-year average spending history in defense, all other programs, the retirement programs and any interest on the debt. If you really take a look at it, even the alternate fiscal scenario, projected spending as a percentage of GDP, is under our 20-year averages.

So I would argue even the $127 trillion dollars is understating it. But let me tell you how it's coming at us. We're always talking about a 10-year budget window, right? Well the first 10 years was about $8 trillion dollars in deficit. The next 10 years was about $31 trillion dollars. The third decade is $88 trillion dollars for a whopping total of $127 trillion dollars. And again, that could be understated.

Now, let me put that in perspective, because let's face it, trillions of dollars is literally incomprehensible. The entire net private asset base of America is $106 trillion dollars. Now the fact that not any American, nobody virtually—it's 0.00000001 percent of Americans understands that information. It's a big reason why we have not forced the federal government, forced elected officials, forced politicians to start grappling with it. Instead, politicians get away with the demagoguery. They get away with the lies.

Now how many times have you heard politicians from both parties say that Social Security is solvent until the year 2031 or 2033? It's almost universal. It's also a complete lie. I think Wally would agree with me as an accountant. You know yeah, we've got a Social Security Trust Fund that's supposed to keep that solvent for the next 20 years. But the trust fund's a fiction. The trust fund holds U.S. government bonds. Okay, that's an asset to that agency within the federal government. But the U.S. government bonds a liability to the federal government, to the Treasury. You consolidate the books of federal government, that nets to zero. OMB (Office of Management and Budget), in their own publication, admits that.

But in seven budget committee hearings, seven, I have laid out those OMB, that publication, those words, calling the trust fund a bookkeeping convention, that transaction nets to zero, only two people, CBO (Congressional Budget Office) Director Elmendorf, Fed. Chair Yellin were honest enough to admit that it has no financial value to the federal government. Treasury Secretary Lew and five OMB directors—or four OMB directors and nominees refuse to admit that truth. As a result, we haven't even begun to address that 30-year problem. And oh, by the way, 30 years, we see a lot—I see a lot of white hair in the room here. We know that's not a very long time period. My little baby girl just turned 31, and that went by like that.

Now, we've got another problem that we talked about all the time, climate change. Now first of all, let me go on the record, I do not deny climate change. We've had it throughout geologic time. I also don't deny that man has some kind of influence on it. I just don't know to what extent. I also don't know to what extent throwing hundreds of billions of dollars at the problem will it have any effect whatsoever. And yet, we're throwing at least tens of billions of dollars to the program. It's on all the news programs. We're talking about it all the time. We're whipping up all this state of fear. And we're spending tens of billions of dollars to address a problem that I don't know if you can even fix.

You know, I come from the State of Wisconsin. Twenty-three thousand years ago, my state was covered in about a 5–6,000 foot thick glacier. I know there were men back then, but there weren't enough men building campfires to cause those glaciers to recede. Something else was happening. You look at the Vostok ice core sample, over 400,000–some thousand years, we've had, I think, five—four or five cycle—or climate cycles, somewhere in the 15–20 degree range. Something else caused that.

Again, I don't deny that man has some kind of effect. I just don't know what it is. I don't know what effect we can possibly have. We had Bjorn Lomborg, somebody I really, really respect, a fellow that realized, you know, you have scarce and limited resources in this world. If you want to solve problems, if you want to alleviate suffering, let's take a look at where those scarce resources are best allocated to benefit the most people. I would argue it is not climate change. That's what he argues as well.

But for some reason, Americans believe Vice-President Al Gore, in 2009, when he said, "In five years, because of climate change, the northern ice cap will be melted." Well, it's 2014, and it still exists. When you look at a lot of people that really promote climate change, yeah, I don't know these, I don't have the exact quote. But let's face it. People promote it say, "I don't care what the science is. This is just such a good opportunity for us to gain control over society." That's basically a paraphrase.

And that's what gets me to my point. Government will solve a problem when the result of the solution allows it to grow. When it allows politicians, when it allows government officials to gain control over Americans' lives. That's climate change. That's why so many people are all about spending tens of billions of dollars to solve climate change, because it gives them control. You know, solving the entitlement problem, our debt and deficit issue, what does that result in? That results in less government. That results in less control.

So now let's take a look at EMP. I am amazed, having come here now, and finally being made aware of this by Dr. Pfaltzgraff and Ambassador Cooper, this was made public in 2004. In 2008, and that's what this is right here, this is the preface of the report, it's talking about the dependence of the U.S. society on the electrical power system, its vulnerability to an EMP attack, coupled by the EMP's particular damage mechanisms, creates the possibility of long-term catastrophic consequences.

If our power grid is down for 12 months, this predicts 90 percent of America would perish. I think that's a problem. [laughter] Now, you know, I've heard a number—I know James Woolsey wrote an article in the *Wall Street Journal*. I don't know the truth of this. I don't know the exact numbers. Supposedly, $2 billion dollars would take us a long way toward protecting our infrastructure here.

When I read that, the first thing I thought about, okay, this was known in 2004, 2008, then we had the great recession. Washington's response was spending, what, $800 billion dollars on a stimulus for what?—Infrastructure projects, right? Why didn't we spend $2 billion dollars in 2009 to at least start addressing this? Could it be because solving this problem doesn't increase government's control over our lives? It's a possibility.

What we have is a huge challenge on our hands. And what I like about the CIPA bill is it just starts that process. You know, it's not even nearly enough. But in order to solve a problem—let me go back to my first part of my remarks—The first step in solving any problem is raising your hand and admitting you have one. And then you have to properly define it.

Now, let me just tell you my little journey here with EMP. It's pretty confusing. I don't know much about it, but I know a whole lot more about it than most Americans. What I'm starting to find out is it is pretty confusing. When I raised the issue with one of my colleagues, I'm not going to identify anybody, the first words out of the person's mouth was, "Yeah, that's just the—That's the transformer lobby. They just want to sell the government some goods." Well, maybe that's a possibility.

So there is going to be all kinds of political pressure, back and forth. What I want to use—and this is my dedication, here is my "ask," then I'll shut up. It's pretty unusual for somebody who's been serving in the U.S. Senate for four years to become a Chairman of a Committee, much less a pretty crucial committee like Homeland Security Governmental Affairs. I'm a business guy. I'm a problem-solver. I want to utilize that committee to hold hearings, not as show trials, but it's really fact-finding missions, to highlight, to describe, to define problems, in a very truthful, in a very honest way.

So what I need is I need people's help. What I'm not going to do, what I don't want to do, is announce a hearing next week, and then all of a sudden call people up, "Hey, can you come testify?" What I want people thinking about right now is, how can we make this case to the American public? How can we do it so we don't do what Al Gore did, and make wild predictions that are just simply proven false after five years? How are we going to do this in a very rational way, so that the people promoting it aren't accused of having tinfoil hats on? How do we get the science out there when, in far too many cases in the past, science has been misused?

So what I'm really asking is I need a lot of help. Start thinking about, how do we define and describe this problem to the American public? What's the best way, in a relatively short time period, because you don't have a whole lot of time in hearings. And you're going to have Senators asking you questions. A lot of them will grandstand. I want to design a committee hearing so we really elicit truth and real information. Because that is the only way we are going to collectively be able to solve this problem and have politicians, have elected officials bring what they can to bear, to actually solve this problem.

So that's what we all need help on. And you guys are the experts here. And you guys have to understand the competing political interests here. And you got to understand how the public is going to perceive industry lobbying for this versus that. So it's going to be a very difficult issue to bring up. It's going to be a very difficult issue to really get the public to admit they have the problem and start working toward the real solutions. And, what I'm finding out, is the solutions are not that easy. They are relatively complex. And Washington does not deal with complexity very well.

So with that, I think I'm out of time. But I'm happy to take a question or two unless you want to give me the hook.

[side remarks]

Dr. BAKER: What are the chances that the CIPA (Critical Infrastructure Protection) Act will pass the Senate in a rapid fashion? Can you maybe project?

SENATOR JOHNSON: Well, it's not going to pass in the lame duck. There's no time for it. Again, get me the information. Let's hold some hearings. And we first have to actually introduce the bill in the Senate. But that's better done in the next Congress anyway. So again, I am very—There's nothing threatening about CIPA Act. Well I guess the third part threatens industry a little bit. We need to make sure that we calm everybody down. But you know, from my standpoint, CIPA is primarily there to highlight the issue, forcing DHS to take a look at this.

And by the way, we had, in our threat assessment hearing, I asked about EMP, asked about it in a secure briefing. And I'll tell you, DHS is not doing much about it. So it would force DHS to start taking a look at that. And again, this is all about getting public awareness of this stuff. So, from my standpoint, the first step is holding the hearings. And if we make an impactful case, we should be able to pass that pretty easily. It just flew through the House, okay.

CHUCK MANTO: One more question from Mr. Popik. Identify yourself—

TOM POPIK: Hi, Tom Popik, Resilient Societies. My group has been involved in the EMP issue for several years now. One question that comes up again and again is the cost factor. I think you indicated that some cost estimates were low some were higher. One of the things that's been proposed is a level of protection which wouldn't be as complete as for the U.S. strategic forces, but enough to produce strategic deterrents. So in other words, a North Korea or Iran would know that we have enough protection that it would make them think very carefully before engaging in an EMP attack. I wonder if you could comment briefly on the politics of deterrents for EMP.

SENATOR JOHNSON: Sure. Yeah. I mean you also have GMD (geomagnetic-induced current) , by the way. So you want to take a look at this whole—the whole thing that affects our electrical grid. I mean you've got missile defense. From my standpoint, again, problem-solver, what I like to see is information. I would like to have the electrical grid described. I would like to see how many of these huge transformers, how long a lead time are these things? Should we be spending the money to have backups? I mean should we put those things in place, get those things shielded?

I would really like to see a plan. And it first starts with describing what the electrical grid is, in a very organized fashion, and then you just start adding up the costs. I mean this is level one. This is kind of base level what we need to do, in terms of protecting ourselves. Here is Level Two. Here is Level Three. So here is two billion. Here is 10 billion. Here is 25 billion. Here is, whatever.

And well we can—We'll use the $2 billion dollar option. So again, it's about information. It's something that is not used very much in Washington DC. We primarily use demagoguery. I'm into, like, real information. That's what I'm asking you folks, is develop the information in an understandable—keep it simple. You know, the old 'KISS' principle, okay? And I think that's the best way of addressing this.

Thank you very much for your involvement.

[applause]

END

InfraGard and High-Impact Threats

John Riggi, Introduced by Supervisory Special Agent Lauren Schuler

https://www.youtube.com/watch?v=2ynwb4jzDvU&feature=youtu.be

Dupont Summit on Friday, December 5, 2014

Supervisory Special Agent Lauren Schuler introduces Mr. John Riggi, Section Chief for the Cyber Division Outreach Section who outlines the role of the National EMP SIG within InfraGard and its unique contributions to sustainable local communities and critical infrastructure protection from high-impact threats. He also comments on the Critical Infrastructure Protection Act, HR 3410.

LAUREN SHULER: Good morning. It's great to see you all again here this year. I'm Lauren Shuler. I work at the FBI in their InfraGard unit, and I am very honored today to introduce my boss, Mr. John Riggi. He has been working for the FBI for 26 years. He currently serves as the Section Chief for the Cyber Division Outreach section where he leads in the development of mission critical partnerships with the private sector, which includes InfraGard.

Previously Mr. Riggi served as the Assistant Special Agent in Charge for the Washington Field Offices Intelligence Division, leading the ground and air surveillance, language and undercover support programs. He spent several, many other years at FBI headquarters in their Counterterrorism Division, Financing Operations section, and serving two years as the Senior FBI Detailee to the CIA Counterterrorism Center. Prior to that he served 16 years as an agent and a supervisor in the New York field office.

I want to especially mention the awards that he has received during his career. He is the recipient of the FBI Director's Award for leading a highly successful classified terrorism financing interdiction program which was responsible for preventing multiple terrorist attacks in a foreign allies nation. Mr. Riggi is also a recipient of the CIA George H.W. Bush Award for Excellence in Counterterrorism, which is the CIA's highest counterterrorism award, for greatly expanding FBI and CIA joint counterterrorism operations and cooperation.

It is my pleasure to introduce John Riggi to speak to you this morning. Thank you.

[Applause]

JOHN RIGGI: Thank you, Lauren, for that introduction. After that I kind of feel old, I think, is my takeaway. So, I want to thank Senator Johnson, of course, Chuck, for all your work here in this group. And my pleasure, again my pleasure and honor to be here with so many distinguished guests in what is truly InfraGard in its finest form, really, citizen volunteers banding together, driven by mutual concern and patriotism, from all walks of life, representing all critical infrastructure sectors, academia, and government to highlight, address, and contribute expertise to the mitigation of a problem that society and, for that matter, government cannot fully address or even comprehend on its own.

You know, late last night I was doing some research on EMP and high-impact threats, and after I calmed down I certainly became concerned on many fronts, but especially in that we in government who remain so necessarily focused on the immediate threats to our national security, such as terrorism and cyberattacks, that we're not sufficiently focused on the looming and potentially existential threat posed by an EMP event, whether it's naturally generated or by man. My personal view became clear to me that we in government and society are devoting insufficient resources, both financial and intellectual, to this issue.

The good news is that you're helping fill that void. And your collective efforts are having an impact in creating awareness and moving government forward toward research and preparedness for an EMP event is demonstrated through yesterday's exercise. It's fantastic. When I saw that agenda of the list of experts that banded together, again, from all levels of government and private sector to try to help develop a solution, to try to become prepared for such an event there is no doubt in my mind that your collective efforts assisted in the passage of the House bill just this past Monday, HR 3410, as the Senator referenced, which will require DHS to include the threat of EMP events in their national planning scenarios and to conduct a public education campaign about the threat of EMP events and authorize research into its prevention and mitigation.

High-impact threats are clearly uncomfortable and difficult to deal with at times, and at times require a level of confidentiality and nondisclosure agreements that InfraGard inherently is well positioned to handle better than general public dialogue. Mr. Manto called this problem pre-traumatic stress syndrome or disaster or bad news avoidance complex, and I know that you'll have a panel discussion on that immediately following. The concern of panicking the public or releasing sensitive information, that has been in the news a lot this past year, ranging from the emergency U.S. Senate hearings or the disclosure of other cyber vulnerabilities in the grid that showed us how an attack on just nine control substations could take down the entire U.S. grid.

So, but that has to be balanced, right, in any public discussion. Low probability, but high-consequence threats to the grid, especially when that discussion revolves around how much resources should be devoted to that problem.

So, but then I even question low probability of the grid coming down, whether it's EMP or some other significant event. So I think many of you probably became aware recently of DHS's bulletin about the black energy malware that was discovered in industrial control systems. So they published a bulletin just this past October and identified a sophisticated malware, a sophisticated malware campaign that has compromised numerous industrial control systems using a variant of the black energy malware that had previously and publicly been associated with the Russian government. Again, that's, I'm not saying that from the FBI perspective. I'm just saying that is in the public realm.

Analysis indicates that this campaign had been ongoing since 2011. Multiple companies that had been working with Industrial Control Systems Cyber Emergency Response Team (ICS CERT) have identified the malware in their systems, so very, very concerning. So this EMP group, I understand, addresses all manners of threats that could have catastrophic long-term threat to our infrastructure. So as the Senator referenced, the EMP Commission in 2008, within the 12 months of a nationwide blackout 90 percent of the population would perish, starvation, disease, and societal breakdown.

So again, the fact that you're there bringing awareness to this tremendously important issue I think only, not only benefits that cause but benefits society in general. So had this information been shared in a sensitive but unclassified environment through InfraGard it might have been managed without going to the public domain pre-maturely. This is also similar to issues that we're facing right now in the Cholera outbreak and the hesitancy of many to simply shelter in place regarding an improvised nuclear device detonation scenario such as dirty bombs, but again you're bringing a lot of this into the public realm.

So, addressing these high-impact threats can motivate the country to be better prepared and locally resilient than they are now. Since most don't bother to be, don't bother to be resilient, expecting that others would come to their rescues, as Mr. Manto mentioned. But if everybody needs rescued then there is no one left, right, to rescue us individually. But by focusing on these high-impact events that could impact the entire country for more than a month it becomes pretty evident that we all need to do more with our own resources locally and to avoid those devastating vicious attacks.

So the EMP SIG organized federal, state, local government, and private sector to work on these issues otherwise left untouched in a way that can foster more citizen community engagement than otherwise possible. It's evidenced by your work with the National Defense University in 2011 and your ongoing work with the National Guard and the National Governor's Association, tremendous. The EMP SIG and InfraGard are attracting some top national level talent in each of the critical infrastructure areas and have volunteered services to their local communities across the country.

Local InfraGard chapters are banding with the National InfraGard membership board. Working groups across sectors are overlapping through its liaison panels and other working groups. Work products that are the examples of the work that you're doing, the great work, are the annual conference that we sponsor here at the Dupont Summit, this meeting with EMP SIG that we hold in conjuncture also with NOAA in Boulder, Colorado. Tabletop exercises that concluded yesterday will be part of the total package of major disaster contingency planning that will cover cyber, EMP, space, weather scenarios, and will be given to the National Guard, DHS, National Geospatial-Intelligence Agency (NGA) so that they, the government, us, we're able to learn and develop locally relevant plans.

This work product is going to be discussed, I understand, later on today by former Assistant Secretary of Defense, Dr. Paul Stockton, at 1:00 pm.

So as a group the EMP SIG facilitates information sharing and that's what InfraGard is about. By citizens banding together locally, experts trying to develop solutions to problems and helping the government in ways that have not been achieved by any other organization, often because of the political or financial conflicts of interest.

So, the InfraGard EMP SIG is strongly supported by the InfraGard National Board, FBI Headquarters, and FBI Coordinators around the country, and is an example of what InfraGard members can accomplish as they work together.

Some of the new programs being launched by the FBI through InfraGard include on how we do outreach in the cyber world with private sector. As I often talk about, when it comes to the cyber threat, the FBI, the government, cannot address that threat alone. Eighty-five percent of the cyber networks in this country, the information networks are owned by private sector, so the information and intelligence related to cyber threats is on private networks. The government, contrary to what Eric Snowden and others may say, we don't see it all. We have pieces of the puzzle related to intelligence and so forth. Private sector has pieces of the puzzle.

And the other challenge that we have when we're facing a cyber-threat against nation state actors or international criminal organizations is that although that evidence and intelligence lies on private sector networks, private sector has no obligation to call the FBI. They have no regulatory obligation to speak with us. They may have some other regulatory obligations to notify perhaps SEC or a financial regulator, but not to the depth, with the depth of information that we need.

So, what we need is those trusted relationships so folks understand when they come to the FBI that this is a mutual problem that we will work together to solve, because we treat them as victims of crime. And with the emphasis being there on the victim we attempt to preserve their confidentiality and work together. They have pieces of the puzzle. We do. We try to put it together to understand what the plans and intentions, capability of our adversaries are.

InfraGard is a platform for that information sharing and for building those trusted relationships. So we're always looking for creative solutions from private sector, because as the Senator stated, often the incentives and efficiency solutions are gained from private sector, not from government, because inherently we are not profit driven or efficiency driven in many instances.

So, as I leave here today and place all my sensitive electronics in shielded metal containers, I would like to quote Congressman Trent Franks who sponsored the bill that recently passed on Monday. Congressman Franks said, "There is a moment in the life of nearly every problem when it is big enough to be seen by reasonable people and still small enough to be addressed. This is our moment." So on behalf of the FBI, myself, and my family, thank you for all your efforts. I appreciate it.

[Applause]

Question from unidentified audience member: I would think there would be somebody out there, a multi-millionaire who would be stepping ahead of the crowd to develop his own, to protect his own infrastructure. Perhaps Mr. Buffet, Mr. Gates, the Apple Corporation, maybe they're ahead of us and they're just not talking about it. Do we know anything about an effort or is there any type of effort like that?

JOHN RIGGI: You know, not to my awareness, but often in the economy there is market forces drive innovation, so I think when there is a particular problem faced by the country and economy again entrepreneurship will probably help fill that void, but I'm not aware of any specific programs such as that.

Post Address Comments by Dr. Earl Motzer

Dr. MOTZER: Thank you very much, Chuck, and good morning, everyone. I'm Earl Motzer and on behalf of the InfraGard National Members Alliance Board of Directors would like to welcome you and to say thank you. You're busy people. We appreciate your taking time out of your busy schedule to come and learn more and share more about the electromagnetic pulse.

Members of our Board are pleased to be a sponsor and we are the private sector side. You have heard from the FBI's side of InfraGard. We are the private sector side. And we're pleased to be a sponsor and sincerely hope that you will have some tangible and easy-to-understand take-homes that can be used to educate others on the importance of being prepared to maintain resilience of our nation's critical infrastructure.

Members of our board also appreciate very much Chuck Manto who is the Chair of our EMP SIG and for planning this program. Thank you, Chuck, for all you're doing, and carry on. I'll get out of your way.

CHUCK MANTO: Thank you very much, sir.

[Applause]

Pre-Traumatic Stress Disorder and High-Impact Events, Maintaining Public Calm and Order

Honorable Dr. Roscoe Bartlett, former U.S. Congressman, MD
Mary D. Lasky—Johns Hopkins Applied Physics Laboratory
Dr. Richard M. Krieg—Krieg Group
Dr. Ben Sheppard—George Washington University

https://www.youtube.com/watch?v=KJAQbh67KeQ&feature=youtu.be

Presented at the Dupont Summit on Friday, December 5, 2014

"Pre-traumatic Stress Disorder and High-Impact Events, Maintaining Public Calm and Order" Introduced by Mrs. Mary Lasky This panel examines reasons why discussions and planning for major disasters are difficult and often avoided. Examples are given including planning accomplished in Howard County, MD for the possible event of a small ground-burst nuclear weapon near the White House and how the public should be encouraged to shelter in place for 24–48 hours rather than try to evacuate through fallout clouds. Research was shown that depict best practice and that responsible dissemination of this information and this type of information does not result in panic.

MARY LASKY, MODERATOR: Thank you, Chuck. It's very nice to be here today to talk about this. We've all heard about post-traumatic stress syndrome. And this session was called the Pre-Traumatic Stress Disorder of High-Impact Events: Maintaining the Public Calm and Order. So anxiety and panic about a high-impact event. Actually, we've seen this. We've seen this really recently. We've seen this in our country when Ebola became the hot issue. It was with one hospital in Dallas, didn't look like it was so prepared. And then two nurses got sick. And the whole country just panicked about, "What is going to happen about Ebola? What are we going to do?" Thank goodness there was a lull, and all of. We have 35 hospitals around the country that are really prepared. They can take care of it. They're all prepared. And we don't need to panic in the future.

There is a new book that has just put out by Judith Rodin, called *The Resilience Dividend.* She talks about resiliency. And she says that resiliency is an individual or community and organization or a natural system to prepare for disruption and recover from shocks and stress, to adapt and grow from disruptive events. As you build resiliency, therefore, you are more able to prevent and mitigate stresses and shocks.

So we do prepare for certain events. We prepare for hurricanes. We have the all-American ShakeOut that was started in California and now has spread out through the whole country, of preparing if there is an earthquake. But there are certain things that we do not discuss. Those are the really big problems. We don't talk about nuclear. We don't talk about biological events. And we don't talk about what if we don't have electrical power for months, or even a year.

Often, we're concerned about not talking about it, because we trust our first responders. They're always going to be there. Everything will be fine. Or, the government doesn't want to talk about it because we'll panic the customers and the citizens. Today we have a panel that has been doing research into this. Dr. Richard Krieg is going to talk about research here in Maryland, about a nuclear attack in Washington DC, telling people how to save lives. Dr. Ben Shepherd will discuss research that he has done with a Start Program at University of Maryland and at George Washington University. Jessica Weiner, who is not able to be here, is the leader of the FEMA (Federal Emergency Management Agency) Nuclear Communication Working Group and has produced an excellent document. Because I'm a member of that working group, I will address her slides. And then we have Congressman Bartlett, who has described the challenges of the EMP. And he will talk about working with Congress and trying to move the EMP agenda.

If we do mention these big events, will people panic? What is the result of our research? So Rich, would you take it away?

[side remarks]

Dr. RICHARD KRIEG: Well, let me tell you why I'm here today. Back in the 1960s, when we were under the regime of mutually assured destruction, President Kennedy put Robert McNamara, head of DoD in charge of the Civil Service function within the government. And every place you looked, cover of *Life Magazine,* spots on television, that Gene Hackman, the actor, his first job was as a—was in one of those infomercials back then. But everywhere you looked, including the sides of buildings, people talked about nuclear risk, fallout, and so on, under a thermal nuclear exchange.

Today, you don't find that. And, as I've looked at this issue over time, part of my background, most of my background is in public health. I served as Commissioner of Health for the City of Chicago. And, as I looked at this as a public health person—I'm not a nuclear physicist or an engineer—we have a concept called "excess death in public health." —And that's pretty much people who could have been saved in a situation, but are not saved, or experience morbidity or mortality.

Most of the modeling that's been done around a nuclear blast in a large city has been done by the National Lab Sandia, principally Lawrence Livermore. And in their modeling, if you take the most benign situation of a fish in the bowl nuclear detonation, the estimates are that about 250,000 people in the National Capitol Region would die if they did not shelter in place. And most people do not know that their muscle memory should be the shelter where they are, particularly in commercial buildings, but also in wood-framed buildings that have a basement, and so on.

So when you look at this issue, and you look at the aspects of fallout of the deterioration of the danger in fallout, which is probably 24–48 hours under that situation, a ground-level blast, where the blast effects are mitigated by the buildings, you're not talking about an aerial strike like Dr. Baker did, but rather where everything in that blast zone goes up two miles, and the nuclear plume imposes a danger. And, if you take that 10 kiloton scenario, that's what the National Labs have looked at in terms of the damage in Washington and the concentric rings of blast zone effects.

Oh, no slides.

Dr. RICHARD KRIEG: Okay. One of the groups that looks at the situation at the threat level is the Nuclear Threat Initiative that former Senator Sam Nunn, who headed up the Committee on—the Armed Services Committee, as well as Henry Kissinger, former Commerce Secretary George Schultz, the three of them set up NTI, which is based in Washington DC, with Senator Nunn as the head. And they monitored the theft of fissile materials around the world. They are very attentive to the statements of terrorist groups. They look at the threat levels imposed by the storage of nuclear facilities. And, when you go up and speak to them, they are very concerned about the situation, both state actors, where the threat level undoubtedly is going up, the instability in Pakistan, the relationship now with Russia, where most of this—most of these—most of the fissile material is stored, as well as Iran, obviously, and North Korea, that it is possible to imagine that these countries would think of this as a strategy. They have thought of this as a strategy. And then, when you look at the non-state actors, most of the terrorist organizations, including al Qaeda, have been very specific in their interest in having building a bomb, or securing a bomb to strike Washington, New York, Chicago, very specifically.

So what we've done is, we've been working in Howard County, Maryland, and with the Maryland Emergency Management Association. And we've had a succession of projects that were designed to alert the local community to this issue. The southern tip of Howard County–Howard County is midway between Washington and Baltimore. The southern flank of the county is 25 miles from the White House. And, as we've worked there, our specific intent was to develop strategies to get the community aware of this issue, and to have them have that muscle memory where they know that they have to quickly get into shelter.

So how we've approached this has been a series of projects. FEMA has been helping us on several of them. We worked with MEMA (Maryland Emergency Management Agency). Back in 2013, we had Brooke Buddemeier, who is the lead person at Lawrence Livermore on this issue. And Brooke goes around to the major cities of the United States and meets with the first responders on this issue. It's done on a low-key basis. It's not done on a public basis. But he's made a number of circuits around the country. We had him come in and speak to about 200 community leaders in Howard County about the threat, about how to respond to this sort of a situation.

Back in March 5, 2013, myself and my colleague Dan Hanfling, who's Head of Emergency Services at Inova Medical System in Virginia, we convened about 300 health and hospital leaders in Howard County at the Applied Physics Laboratory, to talk about this scenario. And it was the first time that health people had sat down to look at the implications of a nuclear blast in the nation's Capital.

And we came up with recommendations. And one of the chief recommendations is that, because the assumption is that the word about sheltering in place will not come down from the federal level. We have to have a highly strategic approach to work with communities, so that that message is getting across. And the best intervention we've had, really, is to go in and, for example, have 100 people representing that community, first responders, nonprofit leaders, health people, educators in the public school system.

We present them with a very realistic scenario, a video that we produce, that shows this blast happening in Washington. It's high quality. It's emotional. It's gripping. It's accurate, scientifically. And then, we give them a pretty quick overview of the science, in terms of that detonation. What can you expect in the aftermath of that? And then we talk about policy planning, practice, programs, and how this event should affect that.

So we've had a number of findings that I think are useful. First, it's important to recognize that there's a very high interest in this topic. When people grasp that they're close to the Capitol or to a large urban center, they really get it. They're not unduly concerned. We don't see evidence, at least panic in talking about it. But they get very engaged. We had one meeting in Howard County where it was scheduled to be an hour and a half. It went to three hours. And we had people sitting down with first responders and talking about what would happen.

There is relief that, if this scenario were to occur, a ground blast, and the thinking is that perhaps this would be comparable to the box cutter in 9/11, where it would be a terrorist organization that had a bomb. But they take it in, in a pickup truck, or a panel truck, or a van, and they would detonate it. We use that assumption, knowing that this is a target-rich environment. You only have to drive past NSA to see that that road is just a few blocks from some of the main buildings. You can't drive in because they have barriers there. But it'd be very easy to imagine this happening. We have Fort Mead, Cyber Command, and all of the defense industries. Mary's organization, Applied Physics Lab, has about 5,000 defense and space researchers there. So we're in a target-rich environment. And we realize that this could happen anyplace. But our assumption was that Washington DC would be first on the list.

What we found with first responders is that they are very interested in this topic. There is a concern that, by focusing on this nuclear attack scenario, that it would offset the all-hazards planning that they stand for. The hierarchical NIMS (National Incident Management System) planning that would be invoked in any type of emergency situation; they see this as kind of a sexy topic that captures people's attention. But, in fact, it's obviously a low impact–a high impact, very low probability event, although I think the risk is increasing.

And what we have found in these meetings that we have is that, if you kind of bring it home by saying, "You know, if you're in a hurricane alley, you do all-hazards planning. But you emphasize hurricanes. If you're in an earthquake zone, you do all-hazards planning. But you look at what happens in an earthquake. If you're near Washington DC or a large city, you need to think about this as a possible incident that would happen. And that argument really prevails. I think they understand that. And many of these first responders have become kind of, in Chuck's words, patriots, in terms of wanting to be involved in this issue and lending their support.

With citizens, we found that there is a very strong interest in the topic. We've gone into the public school system. We have presented this sort of—the information to, for example, 76 system principals, the people who are—who have to stand up the individual school emergency response plans. They got it. And they are—they are making changes in the plan. The most difficult issue that exists here, and I was down in Georgia to Centers for Disease Control and Prevention (CDC) doing a tabletop with the Atlanta Public School System. And you could really see this, is that parents want to go to the school, retrieve their child after this blast, and then go home.

And the safest place is in the school. The safest place, by far, even if there is not a basement, if you're in a gymnasium and there's, you know, space up to that roof, that's— that's—that protects you, because there's fallout there. Fallout are like grains of sand. Again, if it was that situation, the fallout effects would dissipate very rapidly, within 24–48 hours. So the students would have to stay there.

And in our work with the schools, and the reports that we produce, we talk about how to deal with parents, how to have students stay there for 24 hours, what are all the implications with that, how does that mesh with strategic plan of the Department of Education in Maryland, and with the Howard County Public School System?

Again, there is no evidence of anxiety. My strong impression is that the people working on this issue, the public people, are very glad to get these facts. They like having the information. And it's their belief that the public can handle this, that we can talk about this, we should talk about this. Over 98 percent of the participants in these meetings and the evaluations that occurred see this as a very important or an important issue across all of the different projects that we've done. Over 95 percent of the participants say it's essential to inform the public about the need to shelter in place following a nuclear attack, and that I think the most common response we get is that if so many people could die in this situation, and so many people could be saved if they realized they had a shelter, and not do exactly the opposite thing, which is to get in their car and endanger themselves, why is it that the government isn't doing more in this area? That's what the public says. They just don't understand that.

And that goes back to our March 5th findings, where the same issue came up. And we just made the strategic decision that we wanted to work at the local level. And, as we have interacted with the Maryland Emergency Management Agency, that's really been the thrust of what we're thinking about, is to get in, do a rapid intervention. Every community has similar types of information mechanisms. But they also are singular, in terms of who are the people that get involved.

And what we've gone to is my final remark, is to go into a community, give them our drill as far as the video and the science, and then ask them, with low-cost or no-cost basis, okay, 30 community leaders, "How is it that you get this word out in your community? How would you use the library system? Who are the connectors in the community? How in the fairly rapid basis could you get out this message of sheltering in place in a nuclear detonation situation?" We've been very encouraged by the results of that, in terms of what the output has been of these kind of crude strategic information campaigns. So with that I'll stop.

Dr. BEN SHEPHERD: Good morning everyone. I hope you are enjoying the electricity. When we're kind of looking at EMP events, it's important to look at the societal reactions to better prepare, respond, and recover from events. And one of the key features we need to really kind of get to grips with is that the public is not prone to panic. And we've seen that time and time again, across the various natural and manmade disasters.

But, rather than panic, people who experience sudden crises respond actively and consistently as possible, according to their pre-disaster roles. And a lot of social science research can actually back that particular point up. However, what we do see is that people may actually change and modify their behaviors, which may actually undermine the safety and well-being of themselves and those around them. It's not that they're panicking, but that they're sort of changing their behaviors, which may actually undermine the whole sort of management and response process, for example, food hoarding or looking to trying to get as much fuel as possible, actions which may actually undermine and have unintended cascading effects for the wider response.

So, looking at the solar EMP, an interesting feature to this is that we may all have a warning of 18–96 hours, give or take a little bit, which raises the question, what do we actually inform the public and tell the public to do during that time? Particularly as, well we heard from one of the speakers yesterday, that you may only actually get a 10 percent probability of accuracy of the warnings say, "Yep, there will be a major geomagnetic storm." And even a warning of a medium level geomagnetic storm may actually result into a very severe level storm.

So we've got some ambiguity uncertainty around that. And so we have the challenge of, well does their responsibility to inform the public conflict with not causing social and disturbance? And we can't actually look at the earthquake arena, where they're actually putting on to say, "What are the challenges for predictive earthquake warning?" As the science becomes a little bit more accurate, and over the years, we'll have more, maybe, predictability in that area, to some degree, how should you then tailor and customize warning systems accordingly? So we can keep a close eye on what they're doing, to then help inform what we can then try and do in the solar EMP environment.

For both solar and the cyber arena, we are unlikely to see looting by the public. Now this is, you often hear about this in various different quarters. But, from looking at the various responses from the public to various disasters, we don't tend to see that occurring. Indeed, if you look at, say, the August, 2003 blackout in New York, there wasn't really any major looting there. And the blackout extended from Ohio, Toronto to Ottawa, lasted about 14–24 hours, okay, a lot shorter period than a major solar EMP would be.

But, what we saw that's kind of spontaneous humanity, people sharing water, assisting others over the barriers, and even people spontaneously taking on traffic control responsibilities. So there was concern, resolve, and excitement. So that's kind of a good, fairly decent baseline to work with the public, particularly during the earlier part following a disaster, to actually keep them onboard and inform them what actions, say, might take to help better get society through the event itself.

However, where we may see preexisting social tensions, that may actually then lead to or contribute to civil disturbance. And we saw that actually in the 1977 blackout in New York, where there was arson, looting, of mainly in the low-income areas. And it soared to actually very rapid and significant economic decline in those areas meant that the looting and the violence were in response to the economic suffering and hopelessness that was seen in those particular communities.

But Mayor Bloomberg in 2003 said actually, that New York of 2003 is a very different place to New York in 1977. And I think that actually had a significant impact on how the city responded to the blackout in 2003 versus 1977. And, as other areas of research may suggest that, if you've got underlying social tensions, you put on top of that a major event, a disaster, then those social tensions could then start to come up, up to the fore.

So, what we may see from a major event is that individuals, families, and neighbors will work in small groups, largely unsupervised to respond to the disaster, and use whatever tools available to them, including the acquisition of resources from local businesses to achieve their goals. Not necessarily purchasing them, but they may go in and say, "Well, we need these essential equipment. There's nobody there that will help us to get more resiliency in our community."

That sometimes can be mistakenly seen as looting. Whereas actually, well, here are some individuals who are actually trying to make their way to say, "How can we prudently ensure more resiliency within our society, given the considerable degradation there?" And so that's another kind of feature, say, "Well, what is really a kind of looting for the purpose of looting for self-serving," to that, "Well, we're for general survival." Obviously, someone trying to take off a big flat screen TV following a blackout is clearly not doing it for survival purposes. But, if we're looking at, say, hardware stores, and shops and things, then that can be seen otherwise. But, it's best that's done in a structured way in which there are some authorities to be able to better help and facilitate the process where possible, given, let's say, electronic transactions may not be possible at that particular time.

But public responses to earthquakes show that we have this self-reported behavior of persons who experience the earthquakes on the whole controlled, rational and adaptive and varied. And particularly say, look at the Loma Prieta earthquake in 1989, or Northridge in 1994, and there's been a host of others as well in that.

And also, with a major solar event, a key feature that we need to keep in mind, and with a cyber attack as well, is that there will not be the direct physical damage and casualties that we'd experience from a kinetic attack, say an IED (Institute of Economic Development) event. So we have that kind of upside to this.

So, when we kind of look at what we should do to these areas, it's probably worth asking the question, what are the upsides of this narrative we're dealing with? What is still working? What can we do? What is the most effective way we could better exploit these functions that are still operating? Rather than just focusing on what's not working, what's not getting in there, because that way, if we turn it the other way around, we'd probably then be able to find some more innovative solutions to the various different scenarios we're working with.

Now communications is going to be essential before, during and after the events. Now, you know, we've heard a lot about, okay, the public is not prone to panic. And actually, they're kind of, well, they're very receptive to information on what they might need to do to try to prepare for events. Well, should an event actually take place, some of the essential attributes could include communicating a basic survival guide to either, say, cell phones or computers, while the system is still there but maybe gradually degrading, for a major solar EMP event, not all power go out at once. It would probably likely be a gradual degradation as the electricity or the high transformer, voltage transformers gradually go out of action.

So there's going to be that window of opportunity to pump out this information of what can you do, as a society and individuals, for the next even couple of months, for the basic survival, say, collecting rainwater, storing and cooking meat, and some advice on that, and dealing with sewage and trash and likewise. And also, to make it clear that this is a nationwide event, not a local event.

 And so this would prevent, say, individuals getting in their cars and looking to keep driving until they find the lights are on. This was a good point made by a group yesterday. And that would then reduce the potential of individuals and families clogging up the roads needlessly, running out of petrol and being stranded in the middle of West Virginia or wherever.

Other kind of communication aspects is that, to convey this may last for several months. And all these features will then actually provide some degree of hope and sense that someone is out there to try and respond and deal with this particular situation. And then the sense of hope is really going to be important to get the community and communities through such a situation.

For a nuclear event, just a couple brief words on this. OK, it is a naturally—a no-notice event. But we're also going to see, maybe, increased dread from that, from the survivors, those not in the immediate impact area. Those who are in more of the immediate impact area, the concerns are more likely to be physical safety, essential food, water, shelter, while the EMP is fully more like to be third to that.

But common to both of them, just quickly finishing on this, is that we need to look at what's best for society, not for individuals, and look at how we can repurpose existing assets to better prepare and respond to these various events.

CONGRESSMAN ROSCOE BARTLETT: I've been asked to comment briefly on my experiences on the Hill, with my legislation, of course, that set in place the EMP Commission. I wanted to set up that Commission, and the Committee didn't want to do it. Probably two thirds of the members of the Armed Services Committee had never heard of EMP. And they thought it was frivolous, and they didn't want to set it up. The principal staff didn't want to set it up. There was one person on the staff that was supporting me. And we brought up legislation to set up the Committee.

And I was told that, "You can't do that, because it might require sequential referral." That means that other committees would have to look at it. That would slow down the bill, and we wouldn't get it out. We had to get the bill out, so you can't do it. I said, "What other committees would be involved?" And so they told me. And I went to the Chair of the committees, and I got a letter waiving jurisdiction. So I took those letters back to the staff and the Chair of the committee. Then they had no defense, and so we got the EMP Commission set up.

When I talked to my colleagues about this, there were two different responses. One was that this isn't going to happen. If something is too good to be true, it probably is too good to be true. This seemed too bad to be true. You mean electricity without—no electricity for a year? Maybe 90 percent of our population gone? Nonsense. This can't happen. And so, they just walked away from it.

Others bought the threat. "Gee, we really need to do something about that, don't we?" But then the tyranny of the urgent took over. The urgent always takes precedence over the important. The thing you got to do today you do today. And the other, you'll do tomorrow. But the sun comes up tomorrow, and what do you know? It's today. So tomorrow never comes. And that's what happened on the Hill.

How many of you have read the book *No Blade of Grass*? Anybody read the book? Okay. The Library of Congress has it I know, because one of my staff went and got the book and read it. It's a story that's relevant here. It's a story of our world when the virus attacks all grass-related plants. And they all die. If you think about that, that's rice, that's corn, that's barley, it's all of the grains that we eat.

It was in Europe, but it hadn't come to America yet. And England was making due with shipments of wheat from Canada. And then they got a notice from Canada, "Gee, the virus has shown up here. And I'm sorry, we have to keep our wheat supply here, because we'll need it here. So no more wheat for England."

So, what England decided to do about this was to have an air raid drill, and bottle up their people in their cities. And then they were going to drop a nuclear weapon on each of those cities, because that was the kindest thing to do. There was no way going to be enough food to keep everybody alive. And, rather than having the people slowly starve to death, they would solve the problem this way. Let's hope we don't come to that kind of a solution relative to EMP.

How many of you here are more than 88 years old? [laughter] Gosh, I'm the oldest person here. I remember, very well, the Cold War. And everybody was involved in that war. Everywhere you went, every public building, there were brochures there to pick up what you should be doing to prepare for that event. We had drills in schools. I worked at NIH (National Institutes of Health). We had a drill there, the clinical center was going to be a big 500-bed hospital. And then the thermonuclear weapon was developed, and we might be a part of the fireball. So we stopped having drills when that happened.

But, you know, I think that people felt better about that, because they were doing something. And they had the feeling that, "Gee, maybe I could make it through this, because I'm storing some food. I'm storing some water. I have a fallout shelter. I worked for IBM (International Business Machines Corporation)." They were loaning their people money, interest-free, to build a backyard shelter there. Everybody was involved. People were doing something about that. I think one of the things we can do to avoid this pre-traumatic stress is to have people involved. It's not hopeless. I'm doing something. Gee, maybe I'll make it through that if I'm prepared. So I think that we need to get people involved.

I rode down here this morning thinking about this. And I looked out there, and I saw all of those big, tall buildings with apartments in them. What in the heck are they going to do to prepare for a year without electricity? I'm going to be interested in what solutions others have for this problem. Then I thought about, what is it, 50 million people on food stamps? How in the heck can they prepare? Everybody ought to have a year's supply of food. That's not hard.

I was in Sam's Club the other day, and there was 50 pounds of rice there, and it went up. I think it was $17 dollars. Now a couple of years ago, it was $16.44, 50 pounds of rice. That's a heck of a lot of rice, isn't it? And there are other things similarly cheap that you could buy. It's pretty darn easy to put by a year's supply of food. But how are those 50 million people on food stamps going to do that?

We face a huge problem here. By the way, one other thing that makes me feel really good is that you're here. Because when I started talking about this, the first person I called was Tom Clancy when I learned about this. Tom is a good supporter. He did fundraisers for me. And he had an EMP scenario in one of his books. So I knew he had done good research. And he said, "Gee, if you read my book, you know all I know. But let me refer you to the smartest man hired by the U.S. government." Gee, that's a tall order. The government hires a lot of people.

To his view, that was a Dr. Lowell Wood from Lawrence Livermore. This was pre-cell phone. Remember the old pagers, up to the satellite and back? And I paged Lowell Wood, I thought he was in Lawrence Livermore in California. He happened to be in Washington, within an hour sitting in my office.

This is a huge, huge problem. It just defies imagination that we're not really doing something meaningful about the problem. We're just barely nibbling at the edges of this problem. But the next panel, I think, will further explore this. But I'm really pleased that you're here, because now this is not just Roscoe Bartlett up here, I went to the floor probably half a dozen times and spoke for an hour on EMP, with a lot of charts and so forth.

Now, Roscoe Bartlett now is represented by hundreds of thousands of people who know something about this. And my hope that we will do something meaningful about it is increased every time I come to one of these events. Thank you very much.

[applause]

MODERATOR MARY LASKY: Thank you. Thank you, panel. That was very interesting. I think that we have time for one question. All right.

Q: Unidentified off-mike audience member asking question about book on preparedness for an improvised nuclear device: Though some of it is outdated, the majority of it is very, very applicable to what you were speaking of now. It is a very lengthy document …

MARY LASKY MODERATOR: Let me try to answer that, because Jessica Weiner, who is not here because she's ill, leads the FEMA Working Group on Communication. And that working group has produced a document that really goes into the current thinking about how to prepare for a nuclear attack and also how to prepare for a situation that would happen at a nuclear power plant.

And so those two documents are out on the web. So, if you query "nuclear" and go into FEMA, you should be able to pick up those two documents. And those documents are really, they're going around and talking to first responders, so that there would be consistent messaging that would come out as that. Sheltering in place and get inside, stay inside, and stay informed, is the kind of current guidelines of what you ought to be doing. So thank you.

END

Federal, State, and Utility Plans for Grid Protection

Honorable Roscoe Bartlett, former U.S. Congressman, MD

Dr. Chris Beck, Vice President, Electric Infrastructure Security Council

Senator Bryce Reeves, Virginia Commonwealth Senate

Honorable Andrea Boland, Outgoing Maine State Representative

https://youtu.be/gu_ebnF_AA0

Presented at the Dupont Summit on Friday, December 5, 2014

This panel reviews and compares the activities and legislation passed in the state of Maine and Virginia in light of slow progress at the federal level. Encouragement of volunteer activities among utilities was discussed through implementation of the EIS "E-Pro Handbook."

CHUCK MANTO *(introducing panel):* If the next panel can join us, as we come up, let me read off who they are. I know folks are wanting to be proper and everything. Dr. Chris Beck, Senator Bryce Reeves, Honorable Andrea Boland. What this panel is going to do is look at what the states are doing on grid protection with some amazing work that's underway now in Virginia and Maine, two different states that are very different from each other, but at the forefront of wrestling with these very difficult issues.

Congressman Bartlett already mentioned a little bit about what his difficulty was in trying to get the feds to do anything about this issue, with the CIPA Act now just passing the House, he might have an additional word to say. (By the way, I forgot to warn the next panel—[*laughter as a chair partially slips through a crack on the stage*]—We had an incident where a fellow from the FBI almost attacked the Senator. If somebody is going to accuse me of trying to do a stunt for public relations—it's not the case. It was just the way the venue happened to work out.)

And then, when they finish, Dr. Chris Beck will take a moment just to talk about what folks are doing on a volunteer basis within the utility industry, to begin to address these issues as well. So we're going to have different perspectives. The federal perspective, just in brief, at the tail end, what the utilities folks are doing internationally. But the bulk of the work I would like to have both Senator Bryce Reeves and Honorable Andrea Boland *discuss*. Andrea, I'd like to maybe start with you as soon as Congressman Bartlett has a few words to say, again, you may have an additional thought about what we've been talking about, how it bridges what the states are doing or trying to do.

CONGRESSMAN BARTLETT: Thank you very much. I'm really pleased to be here. Now I was in Congress for 20 years, and I live about 50 miles, exactly 50 miles to the Capitol. And so I commuted from my home. And, for 19½ of those years, I drove back and forth, and I never saw a building or a house on fire. Now I say 19½ years, because in the last six months it was one, a little one, right across the road from our farm out in Frederick.

And, you know, every one of those buildings I'm pretty sure had fire insurance. Now how likely were they to burn? In 19½ years, I never saw one of them on fire. And yet, I'm sure that every one of those, pretty much sure that every one of those had fire insurance. I know when I have a new structure, I can't sleep well unless I have fire insurance on it.

Now there's far less chance that a single house is going to burn than there is that we're going to have a major emergency required in the kind of planning that we're talking about today. Why in the heck don't we get it as a society? Why do you go into a panic mode if you don't have fire insurance on your house, and yet you're perfectly benign you don't have bad dreams or anything about an EMP attack? Without electricity for a year? Can you imagine what happens in a city like this?

How many of you have read *One Second After*? Okay, big show of hands, all right. And that was in a little town. They did real well, only 80 percent of their population died the first year. And New York City, the novel has it, there were 25,000 people alive at the end of that year, out of the many millions in New York City. Why in the heck aren't we doing something about this? You know, unless everybody does it, that is, protects the grid, nobody is going to do it. Because as soon as you do it, your cost of production is going to go up. And then you're not going to be able to sell your electricity. Somebody else is going to sell it. So, unless everybody does it, nobody is going to do it.

But I was probably the least supporter of big government in the whole Congress, you know. I'm Thomas Jefferson. The role of government is to keep people from hurting each other and otherwise leave you free to pursue your goals. Conrad Burns said that all his people in Montana wanted the government to do was to protect the shores and keep the mail running. Other than that, they didn't need the federal government.

You know, if everybody in the federal government was raptured, how long would it take you to notice they were gone? [laughter] Now, if you wouldn't miss them, why in the heck are you paying for? Huh? Does it make any sense? No it really doesn't, doesn't make any sense. And then they tell you, it would cost too much. Now, if you don't do it, and it occurs, and 90 percent of our people die, even thought it means the end of life as we know it, if we aren't prepared for this and it happens. How in the heck can it cost too much? I don't understand that. Help me understand that.

Just in closing, every one of you needs to be involved at three levels. What are you and your family doing to prepare for this? Don't count on the federal government to solve this problem. They ain't going to do it. Then you need to be involved at some community level, the little town you're in, the club that you're in, the church that you're in, what can you do collectively? And then we need to be involved at the national level. If you aren't involved in all three of those levels, please get involved in all three of those levels. If you have a wife and kids, you know your first responsibility is to them. Thank you for convening this. I'm really pleased to be here.

[applause]

ANDREA BOLAND: Hello. I'm Andrea Boland, newly not a state legislator. So I'm kind of enjoying the freedom of being able to do what I would like to do, and additional things I'd like to do to support this cause. And I'd like to celebrate Congressman Roscoe Bartlett who's been a pioneer and a great leader throughout the country for all of us. And I'd also like to applaud you for being able to get those committees to waive jurisdiction. Some of the bills that I have brought, the leadership works hard to find a committee that was sure to kill it. So they referred to that kind of a committee.

In this case, of course, it came to the Energy, Utilities and Technology Committee in Maine when I brought a bill on protecting the grid from electromagnetic pulse. And the bill I brought was simple. It just said, for everything under construction presently, and everything going forward, it will be necessary to design in protections against electromagnetic pulse and solar storms.

And I took—I knew better, but I took that position to just put forward the assumption that the utilities knew what the problem was and would know how to fix it. But, you know, of course I suspected strongly that they didn't. But at least it would give them a chance to step up. I only asked for protections and for present construction. And we had a huge project going on through the state, which I learned from the experts was actually doubling our vulnerability because of the way it was designed. And apparently, the utilities didn't understand that when they put it up, or weren't worried about it. So there were a lot of flags to suggest that we were going to have a conflict with the utilities.

In any event, I just asked for protection for present construction and future construction and not for everything on the grid, because I didn't want to ask too much. I knew it was going to be a challenging subject. And I also wanted to give the committee an opportunity to put their own stamp on it if they wanted to enlarge it, which they did, which was great.

So, what really happened, and try to be brief on this, was that we had fabulous national experts come to Maine, volunteer their time and their expertise, some of whom are in the room now, and gave such overwhelming testimony to the committee that, in just the single public hearing, which caused the committee members to sit practically immovable for almost five hours, is listening to one after the other give their speech.

At the end of that, the Chair of the committee said, "I don't know what we're going to do, but I know we're going to do something." It was terrifically convincing. The utilities only sent their lobbyists. And the lobbyists really didn't know that much except to say, "Well, it's not that big a problem. Don't worry, we got it under control." And, after hearing all this testimony, of course the committee members thought, "Well, I guess we really do need to hear more." Anyway, as it went on, the utilities, more and more, came forward. But it was a display of incredible incompetence, is what I would call it in the end. We had the public hearing that went for nearly five hours. And then we had three follow-up work sessions. And the picture didn't get any better as we went along.

The committee ended up passing this bill as amended to a resolve, which was to tell the Public Utilities Commission to get a study together to bring all this information together, and more, and bring back to the committee the list of the vulnerabilities, their options for correcting it, low, medium, high cost choices, and a framework for putting these different options in place. It was clear they expected something to get done, and just give us a clear picture of what our tools would be.

They gave them the time. Everybody signed off, even the utilities, on the fact that it was a worthwhile thing to do. And, when the time came for them to deliver the report, they had only dealt with solar storms. And they hadn't done that very impressively. And there had been incredible input that it was on an online docket anyone could see, and anyone could contribute to. So that was a little discouraging.

They at least had gotten to the point of finally, at that time, the bill passed in May. And they had—well, it didn't fully pass until June. But anyway, they had until the end of January to complete a report. And they said, "Oh certainly, we'll get it done." So, of course they didn't have it done. And the good part that looked like progress is, at least, they were acknowledging that there was a problem, and they should work on it.

So they asked for an extension. And the extension is coming due this month. They convened a working group. But, in this case, the Public Utilities Commission kicked the job over to Central Maine Power Company, the big power company, and asked them to include others to be part of it. And others did participate. The Foundation for Resilient Societies that's here in the room has been a particularly loyal and close and disciplined partner for all of this. They gave them incredible information to work with. And still, they have not addressed electromagnetic pulse, only solar storms. And they're trying to say they're not going to.

And also, with the help of private research and development business in Primus, they've had great support also, as far as helping them understand some options that they could do. And it's a struggle that isn't finished. They've got everything they need to work with. They have their own data that they've generated from solar storms and the effects on the grid. And their first draft proposal included none of it. It included a Natural Environment Research Council (NERC) scenario which understated greatly the likelihood of there being a serious event.

So that's where we are right now. And I, if there's anything here that I wanted to—well, I guess that's the high point of it. The good thing about it is the committee got it, it was clear. They heard the speakers, and they were convincing. They followed them to hotels and talked with them afterwards. The legislature, the House of Representatives passed the bill unanimously. The Senate passed it 32-3. And, even though we have a very highly inclined-to-veto governor, he let it pass without his signature, which to me was a huge victory.

So we've got everything in place. I think it's really good legislation. You'll see it in this booklet that Chuck is handing out for you. It just is a great guideline for anyone else going forward. Just give us the facts, well sort them out. Give us what it will cost. Very inexpensive. We figured it would really cost, at the most, maybe $2.50 (*per person*) a year for five years. As one person commented, it's the cost of a movie ticket to get this enough protection in place that we would survive a major event. And it could cost a lot less than that, depending on how the ISO (International Organization for Standardization) does it. It could be just 25 cents or 35 cents a year. And even in Maine, we can afford that.

So I want to just tell you that we continue to fight really hard for this. And I must say, Congressman Bartlett asked, how can it cost too much? And I have to tell you, Congressman Bartlett, if you would like to come to Maine, Central Maine Power can give you all sorts of reasons that it could cost too much. Somehow or other, they really don't recognize the benefits which could actually increase their profits if they got this thing under control. But they are squirming and diverting attention in every direction they can to try not to do this.

We're going to hear from another state that has an entirely different story to tell, which will make me look totally inept. But anyway, that's the report from Maine.

[applause]

SENATOR BRYCE REEVES: Good morning. All is doom and gloom. Merry Christmas. I not only get to deliver some EMP good news, but we only have a $2.4 billion dollar shortfall in our coffers this year to deal with in a few months, in less than a month. So we're dealing with a lot in Virginia, and I'm sure quite a few of you live in Virginia. Let me just tell you where we are with our standing with EMP and why I'm standing at this lectern or podium this morning.

I guess it was last year at some point when Scott Cooper, who is in the audience (I want to thank Scott for that, piling more on my plate), Scott's father Ambassador Hank Cooper and our former TAG (Adjutant General) General Bob Newman came to me with an issue on EMP. And let me tell you my background. At one point in my life I wore a pixilated digital uniform, armor ranger type 1H. And I'm a fixer. So this was an issue that, on one of the commissions I serve on, I serve on the Commission on Military Installation of Defense Activities. Most of you all know what we went through back on 9/11 and how closely tied to the federal government, Virginia is, and especially to the Pentagon and our military establishment.

And, after going through several iterations of Operation Capitol Shield, which was the big scenario we play in Northern Virginia, we found that this issue was relevant. And so last year, I asked for a study. I knew there were several congressional studies out there and a lot of information out there. And I think we had a quasi-terrorist attack in California on the grid. So it went to the forefront. We pushed it out last year in a bipartisan effort. It was signed.

The study is still ongoing. I've got a couple of pieces of legislation we're already going to start with. That actual study, we met with Secretary Moran two or three months ago. He was very receptive, drinking from the fire hose as a newly installed Secretary of Public Safety. But definitely understood the urgency to which we needed to make sure our emergency management personnel included us in as part of our preparedness.

So I have two bills I'm going to drop Monday that have come out of the commission now. I wasn't sure I was going to have legislation. So I'll have continuing legislation. One is, as we conduct our statewide drills, we're going to include EMP on there, almost somewhere as tabletop exercises some of you all were involved in yesterday. And we'll make that a formal process. And the other part is, to make sure any recommendations or private entities will be included in that.

I will say, in Virginia, we're probably a little bit different. I know our Dominion representative is here, one of them is here. And someone had touched on proprietary information. We've kept a lot of that confidential. But I think there's a collective effort in Virginia, probably more so out of all the 19 military installations that we have in our proximity to the federal government, to make sure that we do this right.

And so we're going to continue those efforts. I haven't run into a whole lot of resistance. At first people thought I would be wearing an aluminum hat or something when you come up with this issue. But I think someone mentioned earlier about, "Show us the facts. Show us the data." It's reasonable. It's prudent. And it's something we need to do. Failure to plan means you're planning to fail. And we don't want to do that. I don't want to look someone in the eye and say, "Why didn't you tell me, when we lose our power because of an event, we're going to have to do this?"

So the ideas of shelter-in-place and all those things, we're going to get that information out. You know, there was another panelist that was up here. They were talking about a scenario they ran with regard to a pandemic and Ebola. And we have all these big threats that we have to face each and every day. And I think it's our job, especially sitting in the legislature, to do the very best that we can do to make sure the constituency that I represent, and the folks in the Commonwealth of Virginia are prepared for that. The reality is, is that our proximity to the nation's capital makes us a prime target. And so we're going to move forward quickly as we can.

And I want to thank you all. I didn't even know what InfraGard was until you guys brought it to me. And it's nice. Part of the challenge is communication. Communication from the local governments to the State representatives, state representatives to their Congressmen, and down the chain of command as well. And so we're going to continue to work on that effort. And it takes money. And money is a challenge. It's a challenge for us this year.

You know, I've spent the whole week going to various committees, from health services organizations, local planning committees with their hands out. And that is a challenge for us. So, as we move forward, this is something that we know is definitely on our radar. It's high in the queue, as far as order of fixing. And I can't thank you guys enough for being here. And I think we'll take questions at the end. Thank you.

[applause]

Dr. CHRIS BECK: Thank you very much. I'd like to thank Chuck and the EMP SIG Policy Studies Organization for this meeting. I'd especially like to thank all of my fellow panelists who have been leaders on this issue, Roscoe Bartlett, for several years, and the leadership shown here by Ms. Boland and Bryce are a reminder that this problem is real, but it needs to be addressed on several levels.

And the level I'm going to talk about today is mostly about industry-led and voluntary efforts, as opposed to legislative or regulatory. I'm Chris Beck from the Electric Infrastructure Security Council. And we are a 501.c.3 nonprofit, focused basically on education and information sharing and a convening, a meeting function, which was initially focused very tightly on electromagnetic threats, being GMD (geomagnetic disturbance) and EMP. It's broadened a little bit to what we call black sky hazards. And those are hazards, again, focusing on the electric grid that could cause extended wide duration power outages.

We have these annual meetings that we call summits. And we started those, mostly at the government level, but they've expanded to include industry. And, as the panelists here, and also previous panelists said, a finding of those discussions was that actionable information is really necessary to move things forward. And so what we've tried to do is put together some of that information that we hope will be helpful to all the stakeholders involved. That includes the electric power sector. It includes federal, state, local governments. It also includes the NGO (nongovernmental organization) community.

And we put together a survey report last year called The International Electric Grid Protection Report. And it looked at countries that have expertise, mostly in GMD, such as the Scandinavian countries. But there were 11 countries throughout the world that we looked at. And there are, again, solutions out there. There are certain types of transformers that are used. There are certain grounding configurations that are good to use, etcetera. And so we tried to share that. But that was a sort of high level survey report. We wanted to put more—a finer point on it, so that the next phase is what we called the Electric Grid Protection or E-PRO Handbook that we've just completed the first draft of.

And this focuses on, it's much more as a—we call it the Handbook rather than the Report. So we tried to make it as sort of actionable document that, in the parlance of disaster discussions, goes from left of boom to right of boom. And that means there's a discussion of these various black sky hazards, EMP and GMP, a new Madrid fault earthquake, worse than Super Storm Sandy-type of terrestrial weather event, coordinated physical or kinetic or cyber attacks. Those are the kind of spectrum of threats that really fall into this category of widespread electric grid effects.

There are several industry efforts underway led by Edison Electric Institute (EEI), every North American Transmission Forum, and others, the federal government side, the FERC–NERC process is looking at GMD events as well as coordinated physical attacks. The Department of Energy has led some initiatives on research and also coordination and planning. We just heard a couple state level efforts.

And also, the NGOs are interested. And the reason I think this is important is, in the handbook, we have—so I talked a little bit about the threats. And we lay out, what are these threats? There's a handful of them. Still left of boom, we talked about electric grid mitigations. And there are certain ways that the grid can be made tougher, both through changes in hardware, and also through operational measures. And so we lay out a menu of options that are known, that have been tried, that are in use in certain places, but not widely shared. And we hope that that sharing of the ability to see, "Well here are some options." And we really tried to also focus on cost effectiveness of those options.

So Senator Johnson talked earlier this morning, talked about, "Well how much can I get for this much money? And how much can I get for this much money?" So we lay out a three-level scenario of minimum, intermediate, and comprehensive EMP protection and what those would look like, basically has to do with focusing on which nodes are most critical. Certain things, your black start critical cranking paths, you can go beyond that. You can harden additional things. That's on the mitigation side.

And, on the response side, what's important is, as society gets more and more electrified, when we have a disaster, typically what happens, say, I'll use the Red Cross as an example. Red Cross shows up, there's been an earthquake or a hurricane. Society is disrupted. Power outage is part of that. But they come in, they set up a feeding and medical care station. They're directly addressing citizens who are in distress, who are injured, and who are hungry. And somewhere over there, hopefully the power company is getting the power back on.

But as we become more and more reliant on power, the power outage is actually one of the key elements of the disaster. The longer the power is out, that's actually one of the biggest problems. And so we are trying to get a more comprehensive discussion between the power companies and the federal and state governments and the NGOs to say, how can we all work together to get the power back on faster? Maybe the Red Cross actually needs some direct support to the power companies to help that process. And so, instead of a field hospital, you actually get the power back on to the real hospital in a more expedited manner. So that's the idea of the response side. And, as I said, it's a compendium of actionable items.
It grew out of some discussions among industry groups, among the NGO community, and of the state level, which was hosted by the National Governors Association. So it's, as our role at EIS, we're kind of an aggregator. So none of us sat down at the EIS and thought up some stuff. We all drew this from experts. And we're having a reconvening of those experts in a couple of weeks to present our handbook to them for their review, their approval, make sure we got everything right. But we hope, moving forward, that that is not just a one-off, the plan is for that to have—for it to be an ongoing process, and that we can get that information out so that the responsible authorities and also the power companies and other critical infrastructure owners and operators actually have some very actionable information that they can use to help address this problem.

[applause]

CHUCK MANTO: So, the next few minutes, what I'd like to be able to do is see what questions Maine may have for Virginia, and what questions Virginia might have for Maine. And as starters, I think we have a microphone out there, too. So you get the first question, Andrea, if you're ready. Or I have somebody who's going to be recognized out there with a question. But panelists go first. So how about questions for each other. Yeah, share the microphone there. Congressman Bartlett has a question too. I think you each have a microphone that should work.

CONGRESSMAN BARTLETT: There are two ways that terrorists could bring down the grid. One is through packing a cyber attack. And the other is through a physical damage to our substations. I thought there were 13 critical ones. After the California experience, we told the world there were 11. I don't know how easy it is to find out which 11 they are. But, if we are prepared for EMP attack, how well does that prepare us for these two kinds of terrorist attacks? Or is this another kind of preparation for the cyber attack and the physical attack?

CHUCK MANTO: I'll be glad to give a quick response. But if you guys want to answer it first—But I'll be glad to give a response to that initially. But go ahead.

Sen. BRYCE REEVES: I'll be happy to try to address some of that with, as a state level planning, and talking to our VDAM (Virginia Department of Military Affairs) and those folks, you know, when you talk about EMP, that's one small section of a bigger plan. And I briefly talked about a pandemic, or whatever it is. So there's initial planning and preparation that has to go into all of those. Each is defined differently. And you know, you have to have different remedies. But I think overall, for us, is how do we—and I'm going to break this down, because I'm an infantry officer. Shoot, move, and communicate in that environment? Do we use runners? Or are we protected enough?

So for us, right now, Congressman, and I'll just call it like I see it, because Congress inaction for so long, we can't do that at the state level, because the government that's closest to the people is the one that's probably the most responsive. And so we have to be that way. And so we've taken the bull by the horns. And I know exactly what our National Guard capabilities are in this environment. And so, as we get the rest of it in, we'll know where those deficiencies are. But the biggest thing is, I don't even know, until we finalize some of those things and run through a couple scenarios and tabletops. I'm not going to know that we're prepared for that. But the bigger issue is, we didn't even have it on our radar. I think we're a few steps closer.

ANDREA BOLAND: I'd like to add that, in the event of an electromagnetic pulse attack, or even a solar storm, we could be years recovering from it, particularly if we haven't protected the major transformers which cost about $10 million dollars each, and currently may take up to almost a couple of years to get. And, when we're in line with a lot of other people, maybe other countries requesting them, it could be even longer. And in that period of time, we'd lose about, it's estimated maybe 70–90 percent of our population in the first year. It does just huge physical and disruptive damage, I'd say, really even beyond cyber.

CONGRESSMAN BARTLETT: I think that this sequence however it's initiated-- There are surges of electricity. And the big transformers are blown. And we don't even make them in this country. We can't make them. We order them from Germany or South Korea. And if they haven't had a solar storm, they'll make us one by and by, years that we'll be back in business. My question was, if we are prepared for an EMP attack, does that prepare us for a terrorist attack, either a cyber attack or a physical attack?

Dr. CHRIS BECK: I can address that a little bit. So, on the two sides on the preparation versus the response side, the response for various power outages is much more—the Venn diagram for the different types of threats overlaps quite a bit more on the response side than it does on the preparation side. But there is overlap.

So, for example, preparing for EMP, let's talk about non-nuclear EMP from a radio device. There are certain preparation mitigation options that have to do with standoff distances, access control, some electromagnetic fencing that can reduce the power spectrum of an IEMI device that would also impede a sniper or someone with a bomb from getting to the critical facility. Preparations for EMP which have to do with the vulnerable— one of the vulnerabilities for EMP, both non-nuclear and nuclear is control systems and computers.

And so protecting those, the protections for those computers are more physical. But there could be some overlap with also just the heightened sensitivity to protecting the control systems from those kinds of attack vectors would also likely lead folks to have a more consciousness about protecting their software, what their Internet connectivity is. If you have a very robust system that's actually air-gapped, it's actually going to be stronger from an EMP perspective, as well as a virus-based cyber security threat, as well.

So there's definitely, it's not a 100 percent overlap Venn diagram. But there's definitely a lot of—there is overlap for all of these different threats. And so looking for cost-effective investments are one where you see a lot of Venn diagram overlap, that's probably the best place to put your security investment.

CHUCK MANTO: Yeah, I have a remark I'll add to the table as well. So if somebody who's got a capability of hacking and the problem with industrial controls, it's a very different game than our typical computer right now. It's far more complex. And we have some great cyber experts, by the way, who outlined that threat in our read-ahead material yesterday for the tabletop exercise, which all of you here and watching this are welcome to have.

But, for example, if a hacker goes in and disrupts something, he might tell, in effect, a piece of equipment to turn on when it should turn off, something as simple as that. Turning something on or off at just the wrong time could itself cause damage, or could itself cause, say, a surge, electrical surge through something that then could even cause damage. If you had electromagnetic pulse protection equipment throughout your system, the same piece of equipment that might protect you from an EMP attack, whether it's local or by a high-altitude nuclear burst, might be able to protect the equipment from something that a hacker might do, or something that might just be accidentally caused.

Some of the things that I've worked with, people I've worked with, for example, in the telecommunications industry, understood that. When they had a rack of computer equipment in the telephone central office, some of that equipment radiated so much electromagnetic interference, they had to put EMP protection equipment from each other, so that the motherboard or the board from one part of the system didn't interfere with the board from another. And there's a whole discipline called electromagnetic compatibility that engineers understand how to do that.

So there's a payoff when you systematically ask the questions like you did. How can we make something help us on a day-to-day basis keep the power cleaner, keep accidents from hurting us, or intentional things from hurting us?

The other thing we have in common is this. I think, as a society, I'd say that we stumbled our way, because it's so effective to use these wonderful things that the utility groups have created for us. It's so inexpensive to just plug something in the wall for five cents a kilowatt hour. And we've maximized efficiency by minimizing resiliency. And one of the things that I would encourage us to do, and I think not many of us have done, is say, let's not put all our eggs in one basket. Let's use these potential problems as motivation to create distributed energy, where there's more energy made at the local bases, more energy storage at the local bases, from the utilities as well as from the users, and figure out a way to work together so that everybody can be more involved, whether you're a hospital and 911 center, a data center, or even a community. So I think that's another common plan.

Now, Senator Bryce, did you have a question of someone else? Because we started off with a question. And maybe you have an opportunity to raise on.

SENATOR BRYCE REEVES: I don't. But I would like to hear from those in the audience, because I know this is a brain trust.

CHUCK MANTO: So we have a couple of microphones. Why don't you come to the front. We have one in the back, and then one in the front. The back hand has been up first, and then the front. And so we got one, two, three.

Q—TERRY HILL: My name is Terry Hill. I'm with the Passive House Institute. And Virginia just redid its energy plan. And one of the issues you just mentioned it about distributed energy. And there's all sorts of talk going on right now about the EPA and reducing the output from the plants. There's now another EPA thing coming up about ozone. Where are the synergies between what this group was interested in and reducing energy costs, so the greenhouse gases? I know people don't want to talk about greenhouse gases. But anyway, there's a lot of synergy between all of these things. And particularly in Virginia, we're interested in putting up a DC micro-grid down in Danville that would do the community self-generation.

CHUCK MANTO: So with the passive solar, you have ways of reducing energy in the first place and becoming more energy efficient and using it locally, as passive solar. And then you have micro grids. Is there a comment about either reducing energy need or generating local power?

Sen. BRYCE REEVES: Let me just comment specifically to Virginia, because that's home for me. So the governor puts out—I think it was 23-page, no, it was 423 pages energy plan. Not all of that's going to get implemented, trust me, because we have politics south of the Potomac as we do North of the Potomac. So there are a lot of things in that plan that I've read. I see eye-to-eye with the governor on a couple of issues.

But let me just say this. As far as EMPs and its effect on the citizens of the Commonwealth and our military's installations and our major industry, this is a non-issue for us, that that aspect of the governor's energy plan, because this clearly falls in the wheelhouse of the public safety. I would be lying to you if I told you I read the governor's 423-page energy plan. I mean I know what I took out of it. I carried a bill last year that would create an emergency management fund for offshore drilling, because we didn't have anything in place if we were to be given the option by the administration to even start offshore drilling again, or even looking at that, because we need royalty money. And we need more income.

So I'm of all of the above, as far as energy goes, personally. But there are various costs to that. And I think some of the utility companies that are in the audience today would tell you that some of those costs are outside the realm of reality for them to produce. So that didn't probably clearly tell you. But some of it's going to be political. I'd be a fool not to tell you that some of that's not going to see the light of day. It'll die in committee.

Hon. ANDREA BOLAND: Some of the work that the Foundation for Resilient Societies has done has shown that we would be very likely to save money by protecting against solar storms and for electromagnetic pulse because of the current operational procedures that are the default for the utility companies and having to react particularly to adverse solar weather is costly in having to buy more expensive energy and different other procedures that they need to take.

But, bottom line was, as we all looked at it, it appeared that the power companies would actually be saving money rather than rolling over some of those extra expenses into the bills of the customers.

CHUCK MANTO: Now, we're overdue for time. So we're going to do two really fast questions. And Curtis will do last, because he's going to—no, you're not starting next, so it doesn't matter who goes. Go ahead. [side remarks] Oh, we have a webinar question too. Okay. Bring me the webinar question forward if you can.

Q: Hi. I'm Curtis Birnbach. I'm the President and Chief Technology Officer of Advanced Fusion Systems. I'm not from Washington, so Congressman Bartlett, I'm going to answer your question simply and directly, okay. You asked. EMP, solar storms, cyber attacks, physical attacks, they are all different. They all require totally different protection philosophies. Simple. End of statement. Oh, and Chris, with no disrespect to you, the standoff distances for IEMI pulses are in tens to hundreds of miles. So forget about standoff. There's no such thing as EM fencing. And I'll discuss more about that later. Thank you.

CHUCK MANTO: We have a question from—The note I might mention too, though, is what happens—and you've heard it from others that this is sort of a complicated thing. As you know, there's lots of different types of EM weapons. And they can range from itty-bitty ones to bigger ones. And people are inventing new stuff all the time. You've been watching, if you're careful, we almost have a mini arms race in the electromagnetic field area.

And so, it would be interesting to see what will develop. And so, in many cases, you may, as Chris said, do triage and figure out, well I may not be able to protect everything equally well. But I'll prioritize and do what I can. It's always better to do something rather than nothing. But it's also good to talk with each other, to have friendly challenges, like, "Oh, you think you're safe now? Let me show you." And so I thought it was interesting to do that.

We had a question, is there a list of alternative electric service providers that are ready to set up off grid services? There's probably not a list. But I know it's an emerging market, where you have energy providers trying to play the role of systems integrators to come up with alternatives. We need to do more to provide lists like that. Maybe Chris, do you have such a list on your website yet?

Dr. CHRIS BECK: Well not yet. That's part of the discussion.

CHUCK MANTO: Okay.

Updates on Space Weather Threats for Power and Communications, Why the 2012 Storm Became News in 2014

https://youtu.be/Gu2MKAvMawg

Presented at the Dupont Summit on Friday, December 5, 2014

Mr. Robert Rutledge, Lead of NOAA Space Weather Forecast Office
Mr. Gale Nordling, CEO, Emprimus

Mr. Bob Rutledge reviews the latest findings from review of space weather data including the near miss of the super solar storm of 2012 and plans of the federal government in response. Mr. Nordling covers private sector investment response for critical infrastructure mitigation of space weather threats to electric power infrastructure and latest test results of the solutions offered by his company and used by a utility in the Midwest.

Robert Rutledge (NOAA Space Weather Prediction Center): Thanks very much, Chuck. So I did get a little comfort. I'm not sure which is less understood and well accepted by many, space weather or EMP, but I do think we can change that. I think we can educate, I think getting together in forums like this will ultimately change that. In the meantime, I do take a little bit of comfort of not being the only guy in the room that people don't fully understand, so thanks for that. [laughter]

So this is also important. We heard high-impact, low frequency. And sometimes we don't have all the right people in the room to speak to both sides of that. I own one piece of that; I own the frequency part. Partnered with my colleagues in this room, we can combine that to assess that impact, to really get that comprehensive view of what do we have our hands on.

And we do live with a lot of uncertainty with the sun. It's our nearest star. It's about 4½ billion years old. I had bad news for everyone, in another 4½ billion years, it will explode into our orbit and burn us up. In the meantime, I'd like to enjoy electricity just like the rest of you, and we have been watching the sun for a very finite portion of its lifetime. So we don't have a full understanding. As we go to talk about extreme events, once in 500-year events, 1,000-year events, even 100-year events, please understand we have a tremendous amount of uncertainty on the tail of the distribution.

I think what we do know for sure, and I think what you could get good agreement on, is that the sun is capable of producing disturbances of size and nature significant enough to cause disruption in the bulk power system.

So we have, if we look back in history, we do have some knowledge of events past, and we have looked at accounts of the aurora. You don't have to look far, you can look back in even texts from the biblical era to see where the skies turned red. It's really hard to do much with that quantitatively. We have the Carrington event in our business from 1859. I'll be honest with you, it's the Katrina of our business, it's all a bit of the Sasquatch. Although unlike Sasquatch, I do know what happened. I do know it was very large, and I don't say that completely jokingly. You see estimates of the likelihood of recurrence of the Carrington vary by a factor of two. Not only is that the statistical uncertainty, that's because we can't agree how big it was within a factor of two. That makes it pretty tough to use for a design environment.

And so the title of this panel and the title of this discussion is how did the July 2012 event become news, become a thing of attention now? And I can tell you in the days following that event, and actually leading up to that event, it had my attention then. And I'll flip through just a few slides. I know I went through this in detail yesterday in the tabletop. And I think I can get away with you guys just believing me by wearing a tie in Boulder, I know that doesn't work in this town, so bear with me as I flip through just a few examples.

But it does, it all happens with sun and we can observe sunspots. And those are the drivers, this is the short version, those are the drivers of the significant space weather. So if we look at solar cycle predictions, I get the questions, I see it all the time, "How big is the solar cycle, how active is it? When is solar maximum?" I'm going to ask you to do two things for me when I finish this. One, forget the term "solar maximum;" and two, forget "solar cycle intensity."

It does matter for overall trends and activity, precision navigation, for example. But when we talk the high-impact, low-frequency events, it kind of goes away. I want you to think of it more like tsunamis or earthquakes, that it's really possible at any time. And why do I say that? So if we look back, again this is just an update on the cycle, skipping that, if we look, here's solar cycle 17 starting in the 1930s, is the top panel. Don't worry about the eye chart, qualitative trends here only, the little arrows, you can see there my super bright green pointer, the arrows where solar maximum was. You can see those are distributed are many years, these are years. These are roughly 11 years per panel.—Many years on each side of that.

So again, I mean that in all seriousness. It doesn't mean that much from a significant event perspective. So these are strong storms. And then if we look at the subset of those storms, that really has the potential to cause impacts to critical infrastructure, you can see them here. So October 2003 down here, the very respectable storm in March of 1989, for example. And you can see some of these are mini-years away from the solar maximum. And trust me on this, they don't correlate very well either with the overall size of the solar cycle in overall intensity. So again, the takeaway is don't worry so much about solar maximum. Don't worry about the size of the solar cycle. Know that it's possible, really, at any time.

It's also true as we look back a couple of hundred years. If we look at the 1859 storm, as I mentioned, the 1921 storm which has been used in studies, which was a very, very strong storm as well, the red line—blue is the overall solar cycle intensity by sun spot number. The red line is the average. Both of those, again, occurred in sun spot cycles that were less than average on activity.

And I used the example yesterday, so my apologies to those that hear this twice, but I would ask you; what's more important, the hurricane season with 15 named storms with none of them significant in making land fall, or the one with four with one cap five that comes on shore on the coast of Florida? So the same is true for space weather.

And I spoke, and I started to speak about the July 2012 event, and this is pretty representative of our business as well. And again, not repeating all of those details. But we had in July of 2012, we had a very large and complex region that we thought had the potential to produce significant activity. For the better part of two weeks, every time I looked at my phone I wondered if right now is it going to be the time that I have to make the call to my friends in the grid, to my friends in the other industries, that we may have the big one on our hands? I did not make that call. We had mid-level activity as this thing made its way across the disk and then it rotated out of view. A few days around the limb, kind of it rotates from left to right and produced a fantastic eruption.

What's different between this event and the many, many times it's happened in the past is we didn't have to speculate about what it might have been. It fortunately passed over a spacecraft that happened to be sitting about where Earth would be in orbit, but ahead of us in orbit by about 120 degrees, so the NASA STEREO-A spacecraft. So it passed through that. We have to measure what was in this cloud, and I think I knew at the time it was spectacular. Many people in my business did, but it took time for that to come together again between the impact and the event itself to say what could have happened, what would have happened if this would have occurred 10 days earlier, two weeks earlier, when Earth would have been right in the line of fire?

I would also tell you that the studies that elaborate on that are worth reading and are worth reading in the details. There's a lot to it. They've also taken examples from August of 1972, for example, and say if you flip the magnetic field and the cloud that that produced, we could have had a storm both bigger than what was postulated for this and even the Carrington event, which I referenced earlier. So again, the takeaways are that I think we can get significant activity really at any time. It's hard to get momentum in these issues, and I'd say the same is true for any high-impact, low-frequency event. A volcano, earthquakes, for example, tsunamis. Until it happens, you don't have people's attention.

I think we're trying to do the right thing here, the scientists are looking at these events and saying we had an event in July of 2012. It had the potential to cause significant consequences. I think that's a little more tangible than trying to look back 150 or 60 years and saying, "Hey, way back when something happened that could have caused these problems today." So this concludes my portion of this and I'll be happy to entertain any questions. I think we're going to let Gail say a few words first.

[applause]

GALE NORDLING (Emprimus): Good morning. Emprimis is a research and development company and it's up to us to look at the highest credible threat and parameters to design equipment to mitigate. You know, in all of this discussion over the last four or five years that I've spent with the utilities in the NERC GMD taskforce and so forth, it almost seems like a new reality is emerging which is a cult versus click. And I'll use that just as an expression to say that the utilities are the click and are people hyping things and oh, the sky is falling, the sky is falling, and so everyone gets lumped into a cult, if you will.

I'm not going to say that there wouldn't be disastrous results if a solar super storm occurred. But I'm here today to show a new perspective on this issue. His colleague, Bill Murtagh at National Oceanic and Atmospheric Administration (NOAA) and Lockheed Martin and Zurich Insurance Company completed and published this summer a new study. They took 11,000 claims occurring over 2000–2010 and they correlated them and they knocked out all the claims that had nothing to do with a solar storm. And they made sure the only claims that they left were those that could not be explained away by any other cause.

The results, and remember 2000–2010 didn't involve a single major solar storm, the 1989 Québec storm has been described as maybe one tenth of a 50- or a 100-year storm. There was no 1989 storm during that period. So during that period, they came up within excess, in the United States, of $2 billion per year in claims due to harmonics generated by low-level solar storms. In fact, the real number maybe three to four billions a year.

Well, what happens when these transformers go into saturation and I don't have time to get into it here, but we looked at different technical papers. Ray Walling a PhD from G.E., looked at it. We had Idaho National Labs look at this issue of harmonics. Luis Marti from Hydro One looked at the issue of harmonics for generator rotors. And every single study points to the same result, which is harmonics causes a great deal of disruption and damage.

But this is no longer a one-in-a-fifty or a one-in-a-hundred-year event. This is now an every year event which should be looked at. The analysis we have looked at on a tentative basis would suggest that the harmonics generated from these transformers at low-level GIC (ground-induced current) levels would violate IEEE standard 519 on both the wholesale and a retail or distribution-level basis. So if that's the issue, and we believe it to be, we are urging the electric utility industry to deal with harmonics especially in the new solar standard and on a day-to-day basis.

In FERC's wholesale contracts, there's a provision that says utilities have to comply with basically 519. The same is true, it may be worded a little differently at every state level, and it's deemed a power quality issue. And so that's one of the things that I just wanted to bring up here, and we are certainly urging the electric utility industry to look at this. I think it's not dealt with as nearly comprehensive as it should be in the new solar benchmark standard. So with that, there's a lot of other things we could talk about, but I think our time is probably short.

CHUCK MANTO: Thank you, Gale. I know we have a microphone. Maybe we do one question. We're going to have time for more questions, and of course there'll be a lot of informal opportunity as well. Curtis, you're going to be ready to come up next, I think, because—yes, Curtis will be up in just a moment or two. Any questions for either of them on this? Yes, we have a question in the back, we have a microphone coming to you in the back row. But wait for a mic. State your name and then your question.

GLENN RHOADES: Yeah, my name is Glenn Rhoades. I'm with Best Buy in Lakewood. And I wanted to ask Robert, I know that Bill Murtagh, who I've known for a little bit of the time, and I know he's on a sabbatical—not sabbatical, but TDY at the White House, my question is as a result of the studies that they're doing right now on that special team, are they going to be publishing results? And do you know what the status is of what they're doing? And if you could enlighten us with that, Robert, I'd really appreciate it.

ROBERT RUTLEDGE: Yeah, absolutely. So, I'll speak loudly and then I'll yell at you when they turn it on.

GLENN RHOADES: Please use an Irish brogue, too.

ROBERT RUTLEDGE: Yeah, I can't do that part. But yeah, so that is an ongoing effort and I think it's important that it does have the attention at the highest levels of government, both for this industry, both for aviation, for example, satellite impacts as well. So I don't believe much has been publicly released. They do meet today at 2:00 to give the initial status. I am participating on one of the working groups to contribute to one or seven or eight goals of essentially what would be the strategic plan or vision, or what task would need to be accomplished to provide comprehensive plans. You'll have the policy that will be developed and then from that, they will derive a space weather action plan. So I think they have a pretty short timeline, on the six-to-nine month order, but really it's just in its infancy. But it's encouraging and I think it's comprehensive and hopefully some good things will come out of it. And it should leverage all parts of both industry, government, academic, across all the elements and all the facets of this business.

CHUCK MANTO: I have one quick question myself before we let the panel go. Harmonics, sounds like some things I've heard when I was doing music. What's the impact of these ongoing daily sort of harmonics that you've talked about that seem to be almost a steady state and then gets crazier at certain times? How can operational procedures address those and do they at all, or is there something else you have to do?

GALE NORDLING: That's a really good question. And there's a couple of answers to that. The simulations that we have done with Power World, first of all, would suggest the current operating procedures from most transmission companies would suggest a reduction of ten percent of power transfers when you get at a certain level. The analysis we've done with power world would suggest as far as a voltage collapse, that does very little to stabilize the system. But what's more important, and I think this is a very good question because of that, it doesn't reduce the harmonics on the system. The harmonics are still generated at very low levels. So if you had a significant solar storm, all the operational procedures in the world would not appear to make any difference in the harmonics that are generated as far as the damage that they caused to customers, etc. So, that would point to you need equipment to reduce the harmonics or to mitigate the harmonics.

CHUCK MANTO: Thank you. And Robert?

ROBERT RUTLEDGE: You know what I think, and I should have mentioned this if I'd had more time, so that's a very interesting comment and it's interesting for me that the space weather and the harmonics and the disturbances are being analyzed. I'm also aware of emerging work that is even looking at correlations between activity and spot pricing, for example. So as we go to look at every facet of this and the economic impacts of this, there's a lot of work being done. I can't speak to all of it or substantiate it, but I think it does need to be done to find out what are the cost impacts, which is what we have been doing.

CHUCK MANTO: Yes, it seems like there's an interesting combination of not only how do you manage your network more efficiently and save money day to day by taking these kinds of proactive management capabilities, which actually gives you better control of power networks, but then if you don't do that well, I think what you're basically saying is people could game the system, anticipate what's going on and manipulate the market in anticipation of things that maybe someone at the utility wouldn't bother to anticipate from spot pricing. Is that sort of what you're implying?

ROBERT RUTLEDGE: Yeah, I think—I'm happy to talk off line and find the study, but I'd seen a study that was for the UK grid, for example, that was a fraction of the power they imported at substantially higher cost and their mitigating procedures changed the fraction of power they were getting imported versus generating locally all in solar storm procedures. So there are cost impacts to taking the beginning actions. And I think it's just interesting to try and get a handle of those and then weigh those against some of the things we're talking about here.

CHUCK MANTO: Thank you very much. Hold it, one quick question and then we're going to go. You have a microphone in the back, because we are going to save about five minutes because we're going to skip one of the cameo presentations. So we'll be back on track, no panic at all. If you have a microphone, you need to identify yourself and then ask your question.

FRANK GAFFNEY: My name is Frank Gaffney. I'm with the Center for Security Policy. We sponsor the Secure the Grid Coalition, we're very keen on the discussions here today. I know Curtis is going to be speaking momentarily, but I just wanted to see if we might invite particularly the Emprimus team to address this question of whether there is some bearing on the cyber-threat protection in the kind of hardware that you're describing, or whether it is only relevant to GMD or EMP?

GALE NORDLING: With our blocking device, we only try to solve the GMD and EMP 3 portions.

CHUCK MANTO: Okay, and we did discuss the E1 issue a little earlier—

FRANK GAFFNEY: Just by way of clarification, if I might, Chuck, to the extent that the cyber-threat manifests itself in impulses that your devices would protect against, would they not be relevant to that as well, or is that completely irrelevant?

GALE NORDLING: Our device is triggered on DC or quasi-DC current and/or harmonics. So, if anything else caused those things to occur, yes that would trip the device and protect the transformer and the value at risk (VAR) swings.

CHUCK MANTO: Thank you very much, and let's give a round of applause.

END

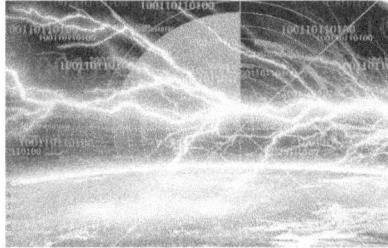

Dispelling the Myths about EMP

https://youtu.be/UZIDDghSlhs

Presented at the Dupont Summit on Friday, December 5, 2014
Mr. Curtis Birnbach, President and Chief Technology Officer of Advanced Fusion Systems

Mr. Curtis Birnbach, President, Advanced Fusion Systems discusses his company's approach to grid protection against manmade EMP and GMD from space weather that includes use of large-scale vacuum tube technology and facilities to conduct EMP testing and simulation. He also expresses his concerns about MIL SPECs for EMP that are later addressed by Dr. George Baker at 4:15.

CURTIS BIRNBACH: Good morning. I'm Curtis Birnbach. I am the President and Chief Technology Officer of Advanced Fusion Systems. Before I start, I would just like to make a quick comment on the last panel. Listen to what these men said. What Dr. Rutledge, who I don't know, said about "It can happen at any moment" is 120 percent correct. And the points that Gail, who I have known for many years, made about harmonics and the problems of harmonics and the inattention of the utilities to the critical nature of the harmonic problem, is also a huge, huge issue. And I applaud you for bringing that forward and making everybody aware of that.

Okay. So some of you know me. And for those of you who do know me, what I'm going to say is not going to surprise you. For those of you who don't know me, I want to say the following. You've probably all heard all manner of statements about the family of electromagnetic threats to the power grid. What you may not realize is that some of these statements are untrue and others self-serving. It is my intention today to set this record straight. This may have the effect of causing some of you to dislike me. So be it. You won't be the first.

There are also various export laws that prevent me from disclosing some critical material because there may or may not be non-U.S. citizens in the office—in the audience. To you, I apologize. But I am constrained by some very severe laws which, in most cases, both of our nations have signed, and we are all bound by them.

That said, there are three basic types of electromagnetic threats: solar coronal mass ejections, nuclear EMP, and non-nuclear EMP, which is also sometimes referred to as IEMI (*intentional electromagnetic interference*).

So, to go a little further on what Dr. Rutledge said about the history of the earth and the sun. About 4.5 billion years ago, the earth and the sun began their dance. Sometime later, specifically August of 1859, there was a massive solar storm which ejected a large plasma ball aimed more or less directly at earth. The effect on the nascent power system and the telegraph system was immediate and dramatic. Northern lights lit the skies almost in the entire Northern Hemisphere.

Since then, there have been several more solar events that have affected the power grid and a number of significant near-misses. We've been studying these effects. But I think it's important to realize that it's naïve to think that, in the prior 4.5 billion years, there were no solar storms, and they've only been around in the last 100 or 150 years.

Unfortunately, there are vested interests that tell us that solar storms can be modeled by a 100-year model. And mathematically, this is totally untenable. And this little graphic I have prepared here explains this. On the top line, we have the creation of the earth running out 4½ billion years. That last line represents a 1 million-year period, slide over here, okay.

The second line spreads out 1 million years. And the last line now represents 1,000 years. The bottom line takes 1,000 years and shows us that, from this whole huge amount of data, this is how much real data we have. And there is no mathematics around that can validly say that a sample of, on the order of three times tenth to the minus sixth, all at the far right end of the graph, can be used as a valid method of analyzing and modeling solar storms. It is pure garbage.

Now, this is a problem. The New York—Let's look at some comparisons. [side remarks]

The New York–New Jersey area was recently devastated by Hurricane Sandy just one year after Hurricane Irene devastated the same region. We were told that Irene was a 100-year storm. So what was Sandy? Where did she come from? Our statistical analysis for hurricanes is about the same database as solar storms.

Now, there's another interesting fact. Let's talk about the U.S. Constitution, the piece of paper down the street, not the ship. Back to our chart. Congress has mandated that the U.S. Constitution be protected against a 500-year event shown in yellow on the same chart. Now, what this says is that a physical piece of paper, which has absolutely no value in protecting the lives and the economy of this country, is to be afforded five times better protection than the power grid. How many people will die if the original copy of the Constitution is destroyed? How many people will starve or die due to the loss of food, medications, life support, water, sewage, if we look at solar storms? Yet NERC, FERC, and other cognizant parties with vested interests are betting all of our lives and the lives of our children and our children's children on this. It is an outrage.

Now, to go one step further, the good folks at the Nuclear Regulatory Commission, who we all know is a very responsible section of the government, recently came out and said that solar storms should be treated on either a 500- or 1,000-year worst-case scenario. So what is it that the Nuclear Regulatory Commission (NRC) knows that the power industry doesn't, and that NERC and FERC think its 100 years is fine for coal plants, but the NRC thinks 1,000 years is appropriate for nuclear plants?

I will also point out that nuclear plants are the only plants that are required to report solar storm events and damage to their facilities from GIC. We have tried desperately to get the information that is held up in NERC, in FERC, at Electric Power Research Institute (EPRI) and other places, information which is public, which these organizations have claimed is classified. Yet they have absolutely no legal right or mandate to classify this information. They are not the Department of Defense. They are not the Department of Homeland Security. They are a bunch of intergovernmental or non-governmental organizations. Yet another outrage.

Solar storms. The Congressional EMP Commission identified about 600 critical nodes on the grid which must be protected. The actual number is somewhat larger. But this gives us an idea of the scale. John Kappenman, who unfortunately is not here today, is perhaps the most knowledgeable person around on the subject of the effects of solar storms on the grid. Having worked on this for over 15 years, being the developer of the neutral blocking concept in 1990 under a contract with EPRI.

And he has estimated that a minimum realistic cost of grid protection just against solar storms is about $4 to $5 billion dollars. And this comes back to the number that Andrea Boland mentioned of 25 or 30 cents per person per year amortized over a five-year period. So this is an interesting concept, because anybody can afford 25 or 30 cents a year. I mean that's just not a realistic premise to think that we can't afford to do this.

The concept of neutral blocking, well John and I are good friends. But, as scientists, we've agreed to disagree on this. I don't believe neutral blocking is a viable concept because, at a point, neutral blocking falls apart. It's a fairly detailed technical discussion. And I'll be glad to have that conversation offline. It's not really the specific topic of this conference.

But, any implementation of protection at all, by any system, no matter how effective it is, is better than none. However, you need to be careful, because if we only partially implement, if, for example, Maine and Virginia are leaders, and they implement GIC protection on their grids, we have what we call the whack-a-mole effect. I'm sure you're all familiar with it. The mole pops up. You hit it. It pops over here, you hit it here. It pops up somewhere else.

GIC is exactly like this. And solar storm protection is really only practical if it is applied on a nationwide or, more correctly, continental basis, because it's just too—too wide an issue. And you can't have—you can't take the energy that wants to pop up here. It's got to go somewhere. So it's going to go somewhere else. And that has to be dealt with.

Some people have referred to protecting against solar storms as EMP protection. This is grossly misleading. It is true that solar storms are very similar in their end result to the E3 portion of a nuclear explosion and the EMP wave form that results from it. But it is not EMP in the classical sense. EMP in the classical sense includes the E1 portion, which is this initial very, very fast rise time, very high transient, that is highly destructive in and of itself.

Advanced Fusion Systems, my company, has patented both domestically and internationally a protective system known as the Field Collapse Method, which, when used in conjunction with its integrated neutral blocking capabilities, is capable of providing 100 percent blocking against the largest conceivable surges from any E1, E2, or E3 sources.

Now, let's take a look at what the largest surge is. Astrophysicists at the University of California have been studying stars of the same class as our sun. They have observed coronal mass ejections as big as 10,000 times the size of the Carrington event. Should such an event hit earth square on, or even as a glancing blow, and particularly if the fields of the coronal mass ejection (CME) cloud and the earth's magnetic field are aligned, well at that point, the atmosphere lights up like a neon lamp and we all die. There have been several mass extinctions that are unexplained. And, while I can't offer any proof that they are due to large CMEs, similarly, no one can offer any proof that they're not. This is opposed to other ones like the KT event and things like that, where it's very clear what caused extinctions.

And so this gets me back, once more, to the 100-year model problem. Do we want to bet our future, our lives, and our futures on a mathematical model which is so specious? Let's talk about nuclear EMP for a while. World War II and the Manhattan Project brought us another class of electromagnetic threat, the electromagnetic pulse. It has been known since 1925, when the noted physicist Arthur Compton postulated that firing a stream of highly energetic photons and electrons into atoms that have a low atomic number caused them to eject a stream of electrons which, in turn, gives rise to a unique class of electromagnetic wave, one with unprecedented ability to damage electrical and electronic devices and systems.

Earlier nuclear testing gave hints of the capability and the extent of the EMP problem. But the technology that was in use at the time, and I'm now talking about 1945 through, say, 1965, was vacuum tubes, electromechanical relays, and mechanical switches. And these are very hard technologies. They're almost impossible to destroy. In fact, in the mid-1990s, the U.S. Intelligence community got hold of a few Russian fighters and were very surprised to find vacuum tubes in their electronics. And everybody said, "Oh, hahaha, the Russians are so primitive."

The Russians weren't primitive, they were smart. They were as smart as foxes, because they knew that vacuum tubes and radar in an aircraft, in a fighter plane, are totally immune to EMP. It was not really appreciated until 1962, when the U.S. conducted a series of high-altitude tests, code named Fishbowl, where hydrogen bombs were set off over Kuwajleen Island. The detonations caused bursts of gamma rays, which interacted with the oxygen and nitrogen in the atmosphere, which in turn released electrons that produced an electromagnetic pulse, that spread for thousands of miles.

EMP knows no political boundaries. As a result of this, street lights in Hawaii were blown out. Radio navigation and communication was disrupted for 18 hours as far away as Australia. Now we're talking about distances of thousands—well, 900 miles from Kuwajleen to Hawaii, and 1,800 or 1,900 to Australia.

So, what can we do to protect this? A realistic cost estimate for EMP protection on a nationwide scale, and this again goes back to the work of the EMP Commission, and the 600 or so critical modes, is somewhere between 40 and 50 billion dollars. Kappenman has done some calculations and has indicated that, should the grid be knocked out for whatever reason, the economic cost of this country is something on the order of a trillion dollars a year. So an insurance policy, as was mentioned by one of the earlier speakers, of $40 or $50 billion as a one-time investment, is a pretty good idea. Typically, you can protect a substation for about 10 percent of the cost of a substation if you're looking for an order of magnitude number of what does it cost to protect transformers or substations.

Okay. Let's now talk about one of my favorite topics, MIL-188-125, the very famous or infamous, depending on your point of view, HEMP (high-altitude electromagnetic pulse) Protection Specification. The U.S. government recognized the cataclysmic nature of the EMP threat and our vulnerability to it. Extensive research evolved the methodology to provide some degree of protection to this threat. Those of you who have heard me speak in the past have heard me rail endlessly against this standard as being inadequate for the protection of the power grid.

Recently, the last six months or so, I've had some detailed conversations with people at the Defense Threat Reduction Agency. And we have come to an understanding, and I have come to a very important understanding. MIL 188-125 is specifically, solely and exclusively designed to offer guidance on how to protect military communications systems, communications systems only, against high-altitude nuclear EMP threats, okay. It is specifically designed to protect command and control functions. It is not designed to protect the power grid. A mistake that I and many other people have made is thinking that 188-125 is a recipe book for how to protect the grid. It offers little or no value. And implementing many of its characteristics is a very, very dangerous path, because it does not take into account the issue of non-nuclear EMP. And non-nuclear EMP is now where I will take this conversation.

Non-nuclear EMP, well those of you who may have seen *Oceans Eleven*—Actually, I got a slide here for this one. Okay. Anyone here see the movie *Oceans Eleven*, George Clooney, all those people? Okay. You may remember there was a wonderful scene where they had a van parked up on a mesa outside of Las Vegas. And they got this really weird looking thing in the back of the truck. And they push a button, and it blacks out Las Vegas.

Well, what you saw there was theoretically correct but visually the work of some absolutely outstanding prop men. This is a system that I built in 2006 as part of a collaborative research and development agreement with the U.S. Army, which, when we tested it out at Picatinny Arsenal, produced 35,000 volts per meter electric field, which is right smack dab in the middle of the EMP energy spectrum. And, more importantly, this particular device punched through the Army's best shielded chamber with only two DB of attenuation. Now that chamber was rated at 120 DB, which is—it's a logarithmic scale. So basically, nothing should get out. And we got out with only a minor loss of power.

Now, this is a real serious problem, because if I can build a system—this cost me about $20,000 dollars to build. If I can build a system like this, so can anyone else who understands the basic underlying principles of physics. And you know, we went through one of the best Faraday cages around, like it wasn't even there. So let's talk about that for a minute. Okay.

The Faraday cage myth. 1836. Michael Faraday observed that a Faraday cage—or a shielded box can operate to block an electrical field within the—Basically, what he did was he built a box that was totally shielded electrically, all conductive, all soldered all together. And he showed that the electric field causes the electric charges within the walls to be distributed. And they cancel the field effect with reference to the interior of the space.

To demonstrate this, he built a room, allowed high-voltage discharges from an electrostatic generator to strike the outside, and had an electroscope inside to show that there was no charge present. The MIL 188 depends strongly on this. When you've heard people today talk about we'll have shielded this and shielded that, depends strongly on this.

Unfortunately, Faraday was unaware of processes such as mirror image waves across boundaries, the what is known as the evanescent field, and also what is known a recent phenomenon that has been discovered known as electromagnetically induced transparency, which is a quantum electrical effect. And these all allow signals to pass through a Faraday shield.

Even worse, there is something known as the inverse Faraday effect, where—and this is a direct quote, because I wanted to make sure I got this definition right. A static magnetization is induced by an external oscillating electrical field at a given frequency, which can be achieved with a high-intensity electromagnetic pulse. The induced magnetization is proportional to the vector product of the fields. Okay.

So, what this tells us is that you can create a pulse, and it will go flying right through this shielded room or this shielded enclosure, as if it it wasn't there. Now, let's talk about stealth for a moment, OK? It's well known the designers of stealth aircraft and other platforms, that even the smallest aperture in a Faraday shield causes a breakdown in the shielding effect. There are some people who have been telling the power grid, the power industry, that you can shield your transformers by wrapping them in a Faraday shield. Well, aside from the fact that this is physically nearly impossible, this slide shows one of the biggest problems.

This is a bushing on a 150 kV transformer at my shop. The hole here is 16 inches in diameter. There's three of them. On the other side, there's three more holes that are eight inches in diameter. To make matter worse, there is a wire that runs right down the middle of this hole, which acts as an antenna, and takes the signal that you're desperately trying to shield and conducts it right into the middle, and happens to be attached to the transformer which you're trying to protect. So if anyone tells you you can wrap a transformer in a Faraday shield, they're full of it. And you can quote me on that. Okay. My notes here said shield acts as a piece of Swiss cheese. But I think you get the picture.

There are also other issues. The shields would have to be large, forgetting about this problem. And there is virtually no space available in substations. They are packed as tightly together as they can be, just by the nature of the design, and the spacing between transformers and devices is governed by the National Electrical Code and ANSI and IEEE specs, so you'd have to rebuild all your substations to do this. Generally speaking, this is not a practical concept. Please, somebody tries to tell you they want to sell you a Faraday shield, tell them you'd rather buy the Brooklyn Bridge. Now, Oh, and my last comment is, other than that, it's a good idea.

Varistors. My topic here is, in italicized type, varistors will save us. Well, as seen in the next view graph, okay, this shows essentially what we have. This black curve is—I'm sorry, the red curve is the classical unclassified EMP wave form. The black curve is the unclassified EMP wave form taken from the EMP Commission Report. Above it you'll see a jagged red line marked "NNEMP." That is data that was taken from some tests that we did with the U.S. Navy down at Pax River, where, with a system that was not much larger than this table, we generated ten to the seventh volts per meter, as measured by the U.S. government.

This is a very distressing issue. And this is why Chris, wherever you are out there, if you're still even here, I know what the near field problems are. And the near field for this device extends out, it extended out way across the Chesapeake and into adjacent states. And we had to have coordination with shipping and aircraft and satellites just to run these tests, because the pulse was so large.

Now, over here, that line is the line that is the 20 nanosecond mark that MIL 188-125 specifies and is what most varistors that are out there, there are some varistors that are a little faster than this. But most varistors are on—have a rise time on that level, so they're slow. They have a large voltage drop. When you run energy through them, because they are semi-conductor devices, which, as John Kappenman so delightfully put it, is like pushing electrons through a rock, you have voltage drop, which causes them to get hot. When they get hot, their resistance changes, and they draw more current, which causes them to get hotter.

And there have actually been failures in large geographical areas of varistor devices from low-level solar storms, not even storms that are big enough to cause a major problem. But, all of a sudden, you just see, you know, in a 30-, 50-, 100-mile, you know, radius area, all the varistors are blowing out. Well, when John told me about this, I thought, well you know, okay, interesting.

So I spoke to one of my consultants, who is a licensed professional electrical engineer and a world class electrical forensics expert and a specialist in transmission generation substation systems. And he immediately knew about it. He'd seen it in Connecticut. So I've got reports, reliable reports coming from Duluth and, what is that, Minnesota, wherever Duluth is. I've got reliable reports coming from Connecticut and from several other places, that this is a possible phenomena.

Now, one of the problems is, that when varistors blow, once you get to a point where you see their capacity, they blow out. They're like fuses. You've got to go out and replace them. And, until you replace them, they are not there and offer you no protection. So that's the problem with varistors.

Now, what are we doing about this? Advanced Fusion Systems has been engaged for the last four years in the construction of a world class manufacturing and test facility designed specifically to build hardware solutions to enable high-efficiency protective devices and also vacuum-based power control electronics. The plant is 250,000 square feet, fully air conditioned. We can build electron tubes up to a million volts in capacity, and capable of continuously supporting currents in the hundreds of thousands of amps. The tubes are thermally insensitive with an operating range, without cooling or any external support, of minus 200 degrees Fahrenheit to plus 1,000 degrees Fahrenheit.

We have an EMP test cell that we built that was built specifically for this issue. We recognize this as—the whole EMP issue as a problem. We recognize there are no good test cells and sort places to do this testing. So we went and built this room which is 80 feet long. Oh, there is the one typo per presentation. It should be 80 feet long, not 80 feet high. It's 80 feet long, 40 feet wide, and 20 feet high. And we have in it—and you can see the interior view on the right—there is a 150 kV transmission line that hooks up to a pair of transformers that are mounted in there. And you can see, my operations director, who is about five-ten, standing in the back, to give you an idea of the scale.

The transformers have the Y connection in the secondary brought out as a fully insulated bushing, so that we can address the issues of harmonics and GIC directly and in a simple form in the laboratory, by merely hooking a power supply up to it without having to go through any great exercise. And, unlike the tests that were run last year at the Idaho National Laboratory, we're not afraid to blow these up. In fact, it is my intention to blow these up. And, in fact, I actually have three of these. And some of you who have been to my shop have seen them—so that I can blow one up and still keep working.

We also have done some work on the area of harmonics, recognizing how important it is. We have built extra instrumentation into the transformer specifically to measure harmonics in the primaries, the delta primaries. So we put a CT *(current transformer)* in there for that. And we have developed some interesting software that allows us to look at the signals coming off of the PTs *(potential transformers)* and separate the even order harmonics from the odd-order harmonics in software, at the touch of a button on the screen.

And this is a very important issue, because the odd-order of harmonics tends to be the damaging ones. And it's really important to be able to understand this. To our knowledge, we are the only people who have such capability. Also, our grounding system is fully instrumented, so we can actually see how these pulses dissipate out into the ground, because that's a very important issue.

We have very state-of-the-art instrumentation. I could go on, but I think I've made my point. And I thank you for giving me this chunk of your time. And I hope that some of you take some of the advice I've given you. Thank you.

[applause]

END

Role of DHS Programs for EMP Protected Emergency Communications and Planning

https://youtu.be/jQi8AA2ZVU4

Presented at the Dupont Summit on Friday, December 5, 2014

Mr. Kevin Briggs, DHS Team Chief, NCCIC (National Cyber Security and
Communications Integration Center)
Dr. George Baker, Professor Emeritus, James Madison University
Mr. Bronius Cikotas, Former Division Chief, Defense Nuclear Agency
Mr. Michael Caruso, ETS-Lindgren, Director, Government
& Specialty Business Development

Panel: "Role of DHS Programs for EMP Protected Emergency Communications and Planning." This panel covers the work within the U.S. Department of Homeland Security to create EMP protection measures for emergency communications systems. Mr. Caruso explained work of the private sector to provide mitigation solutions. Mr. Bron Cikotas, active leader in EMP issues in both DoD and DHS for over four decades, was not able to attend due to serious illness. He subsequently passed from his illness and will be missed by many.

#9—Panel: "Role of DHS Programs for EMP Protected Emergency Communications and Planning"

KEVIN BRIGGS: Okay, thank you. I'll only briefly address first, then. I'm going to try to leave some time for questions at the end here. I'm going to zoom through some of the phenomenology just to level set some of our understanding of these things. Just a quick note on the panel. Unfortunately, Bron Cikotas could not be part of it. He was in the program. You can be praying for him, he went through heart surgery and is recovering. But he's not entirely out of the woods. So for those of you who know Bron, please keep him in your thoughts and prayers.

On the panel we have actually far more distinguished folks than myself here as far as experience and knowledge and so Dr. Baker has asked that I take the lion's share of the time because I'm the DHS guy. I invite you to look at his presentation from yesterday's tabletop and other online things. And then Mike Caruso has over 30 years, I believe, in this business and is one of the top EMP experts in the country for protecting against EMP effects. So I feel very privileged to be not the panel with these two gentlemen.

FELLOW PANELIST COMMENTS: The feeling's mutual.

KEVIN BRIGGS: Okay. Well, thank you. Just a little bit on my background. I've been doing this a whole lot longer than I'd like to admit in the decades time frame now; not as long as them, but I started as a blue suiter back as an Air Force officer, used to be part of the Red Planning Board and we would model EMP as part of that effort. And by the nature of that, I can't really go into details other than we considered it a very, very significant issue.

In addition, I was privileged to be able to help start a program to EMP-harden a lot of the senior leadership communications. So that was my introduction into EMP. Moving forward a few decades as the FEMA readiness division director, we started the IPAWS program. I think many of you are familiar with that. As part of that, when I went to Congress and tried to get justification, we pointed out that the EAS (emergency alert system) alerting network which is—its purpose is to point out when you're having incoming missiles from whatever source, you need to be able to operate through and alert the public. So we started an EMP hardening program at various sites with the help of folks like Bron Cikotas, Dr. Rudaski and many others. I'll speak to that because that's a continuing program.

In addition, currently I serve as the MAC (modeling analysis and continuity) team chief which is modeling analysis and continuity at the NCCIC, we like acronyms, National Cyber Security and Communications Integration Center. So in that role, we model EMP. We are the federal lead for modeling EMP as far as communications impacts and we have a series of mitigation programs.

Let me dive in with a few of—next slide please—so what we have here, as several have pointed out, is that there are three primary sources, or vectors, that create EMP. One is the nuclear weapon detonation, whether it be at high altitude, is very well addressed in the EMP commission hearings and other documents. I commend those to you.

The second, and one I want to mention on the detonation side, EMP commission and most of the effort is focused on the high-altitude side. The EMP from the source region, or for a burst on the deck, that is very significant as well. It's not spatially as great, but if you're looking at the city or the region out to about a hundred miles, even with the low-kiloton weapon, you can have a lot of damage.

Next is extreme solar storms. We've talked a lot about those. I'll just remind folks that weren't there yesterday that back just a couple of years ago, we dodged a CME, a coronal mass ejection bullet. It's more like we dodged a mountain of plasma that was dispersing through the universe that had it gone off about seven days earlier, we would have been in the direct line of fire and we may not have been able to have the meeting today.

The final form, and we were talking about it, some very well here just a few minutes ago, is non-nuclear EMP or radio frequency weapons. There's a whole raft of those that are out there, some on the market. The Russians have been selling the Ronots E for quite some time now; not the latest generation, but it's still quite damaging. You can get them in briefcases. And it goes into frequency realms and types that are much more complicated than some of the previous efforts we've done with just nuclear EMP. So, I'm glad you pointed that out.

As far as types of nuclear EMP, and I'm only going to focus due to time and other things, we're just going to focus on a few of the nuclear EMP effects. We've talked extensively already about high altitude and I'll have some slides later on the source region.

Thank you for doing that. The E1 EMP, I've briefed a lot of folks on this over the years and surprisingly, there is a large number of naysayers out there still that don't believe this effect is a real thing and this is just a community that's trying to—kind of like the self-licking ice cream cone. That is not the case at all. The Russians back in 1962 burst one and during the Cuban missile crisis, amazingly enough, they burst one over one of their neighbors, Kazakhstan and it had extensive damage to their infrastructure. I don't have time to go into it in too much detail, I have some slides that if you have some time later, you can look through some of more detailed components of that.

E2 is not as big of an issue other than E1 knocks out the protection, perhaps, that you typically have with lightning protection. And so E2 is something that we still need to consider. E3 is a very significant issue, especially for your larger yield weapons. And I'll have some more on that here in just a few minutes as well.

The hemp wave form is depicted here. At the high-altitude EMP, you have three phases; the early time, which is up to a millisecond. You have the second, or intermediate time—or excuse me, the first time goes up to the millisecond—or microsecond, sorry—I'll get this right eventually. E1 microsecond; E2, one second, and then the E3 can go on for several minutes.

I'll point out that people sometimes look at E3 and think, "Oh, it's not going to affect anything other than long lines over the continental land mass." Well, you have other systems like undersea cables that surprisingly enough, especially in the shallower areas, you can get significant coupling into those important infrastructures as well.

Here's a very common diagram showing that if you burst at different altitudes, you cover larger areas or smaller areas. At 400 kilometers, you've covered not only the continental United States, but most of the Canadian population, and a good portion of the cities down into Mexico.

This, what we call affectionately the smile diagram from Meditech, shows that when you have a burst, in this case about 400 kilometers above Omaha, that you get good coverage across the United States; again, down into Mexico and up into Canada. And that the focus, or the most intense area, is in the smile area, strangely enough.

The E3 coverage is also very extensive. I'll just touch on the bottom right hand side of this due to time in that if you recall the 1989 electromagnetic storm that was with the Hydro Québec incident. That was roughly one to two volts per kilometer. And what you can see is if you have a one megaton device at the proper altitude, you can cover an area that's close to the size of the contiguous United States.

As far as what is the damage, what happens when you have an EMP, in this case most of these are effects that are demonstrated here are due to the E1 effect. But the trouble is mainly in the tails and so as like one of the Russian simulators has at their site, it's basically the tail's stupid. You've got power lines and you've got data cables, like cat 5 cables, connected into your equipment. And so where you're going to see these antennas, which are like cat 5 cables, a hundred foot cat 5 cable is a really good antenna and it'll couple energy right into the NIC, or the Network Interface Card, and in many cases they would be blown.

Some more examples from our testing. A couple of quotes here. The DTRA, back last year I got this off their site where it says, "EMP would fry any electronics, laptops, sensors, and planes, even a simple cell phone." Now, a qualifier is they're absolute experts in these areas. In fact, here's one of them, I know you consult with them. But hand held devices are generally pretty resilient, whether it be a cell phone or other small device. And where it says it wouldn't kill a single person, it destroys the technology. Well, if you're up in a plane or—without the technology, if the plane isn't functioning, it's a bad scenario, obviously.

The testimony of the Chairman of the EMP commission, I'll just skip through some of these other than the bottom line here where the EMP commission found that within one year of a high-altitude EMP event, you could have upwards of two thirds of the U.S. population at serious risk and perhaps, as he is saying here, they would probably perish from starvation, disease and societal collapse. I'm not saying that's the DHS position, but I am saying we are looking at that very, very seriously.

Here is a redrawn graphic of General Lobro's presentation that he gave back in 1994. It was a great presentation in that it showed what happens when you burst one of these devices over a land mass. And just to summarize, when you have hundreds of kilometers of lines, whether they be power or communications, they found that like on the communications line, every single relay on about a 500-click segment was blown. So you get things like that.

The next one here shows that even—I guess this is dealing with the phone lines—yeah, it shows that all repeater points were blown. And I think it's an important question at the bottom of the slide there, where we asked some of the Russians, "Have you fixed the problems with your communications infrastructure?" And their answer was, "All trunks lines or those going between the major switches are now underground," which was a Ministry of Communications initiative to protect the civilian communications. So when they tested it, they saw major results and they started major programs to fix these problems.

Okay, on the electricity side of the house, on the generation, they found that generators—and these are some of the backup diesels—that the E1 pulse was damaging those, as well as some of the substation electronics. In addition, and this is, again, kind of like the sea water. The E3 will penetrate the ground for a ways. And so where they had about a—I think it was a 1,000 kilometer-long power line that was buried about a meter under the surface, they had extensive damage on that line. And it actually set a power plant on fire. That is an important nuance in that a lot of times, people think of EMP as destroying electronics. They don't understand that EMP also burns down power plants, or can. It can also set fires by the sparking and arcing, especially at the high-frequency type.

From Russian history, and I won't read through all this, I'll just say please take a look at what Kurt Waldheim said back in 1999. I thought it was interesting that a former Soviet ambassador said during the Kosovo crisis that, "If you upset us enough, we are looking at EMP as one of the options." And I know one of the aides that was there said, "And if one weapon doesn't do it, a few more will handle the problem." I know we have, I think, one individual who was actually at that meeting who's perhaps still here. So it's well known by Russian politicians as an option.

Source region. I'll touch on this just very briefly. Sometimes, you'll see where it says this is a one to two kilometer effect. That is very inaccurate. Technically, if you're just saying the technical definition of a source region, yes. But actually the waves keep propagating. And when you do the coupling calculations into the infrastructures, which that's part of what my shop does, you find some interesting results.

Like in this case with a hundred-foot Ethernet cable upset and damaged, what we found is that there could be damage out to about 13 miles and that you could have beyond 40 miles of upset to Ethernet systems. That's not saying that all systems within those rings would be down, but a significant portion could be down. And this one is just from a 10 kt weapon burst at ground level.

So when you look at the physical destruction with the 10 kt, most of the damage is done in that first kilometer. The electronic damage is going out, as you can see here, beyond 10 and up to 40 some-odd miles.

Again, we bought some phones from Wal-Mart, the little power adapters when we pulse those, you would think a cordless phone would be a great thing. And the phone survived fine, but those little wall plugs that do the AC/DC conversion highly vulnerable to the effects. And so we were showing that even with the wrapping of the signal, the prediction was that you would have damage on those wall boards potentially out to beyond 70 miles. Again, that's not everyone, but that would be a significant issue.

FM radio transmission, again out to tens of miles of upset. Actually in this case, over 80 miles with the 10 kt and then damage within about the 10-mile region.

Okay, hardening issues. We need to go beyond the current MIL-Standards. I appreciate you bringing that out. There's a lot of emerging threats. We are working, like was brought out at the introduction, we've been working with the FirstNet community. Dr. Baker here's been helping quite a lot on helping to define other standards that we can use in that domain. If you don't use military standards, I do want to point out there are some very good international standards, at least for nuclear and some other traditional forms of EMP. The IAC and Dr. Rudanski have been doing some great work there. We need massively scalable cost effective approaches, and I'll get into something here in just a—next.

Some conclusions are that the risk of not protecting our critical infrastructure, in this case the communications one, is profound. One EMP burst over the continental United States can disrupt from coast to coast. And a source region electromagnetic pulse (SREMP) burst can disrupt communications within up to a hundred mile area, even with a low-yield weapon. It's relatively inexpensive if you do it up front, and even retrofitting is not prohibitive against the relative risk. We need EMP protection from more than just HEMP, these new emerging radio frequency (RF) weapons, as well as source region EMP need to be addressed. We need massively scalable programs.

We have internal to DHS and the federal community, and we've shared it with a lot of our industry partners as well, we have published an EMP Protection Guidelines for equipment, facilities and data centers. It doesn't get into some of the esoteric, newer emerging threats but it does handle the nuclear and the solar.

Here's just a flavor, it's about a 45-page document right now. We have it broken down into three levels. And the thought here, and this is something I actually heard originally from Chuck years ago, that we need to not just try to use one size fits all where you're using welders to try to handle everything. We don't have enough welders on the planet to handle the EMP problem. You have to come up with other creative solutions.

So on the left-hand side is more the MIL-Standard with some fire suppression and other things added in. Level 2 is where you are using faraday cages against the traditional threats and fire suppressors for buildings. It's not much good if you have an EMP steel room in a building that burns down. You have to do it holistically.

Level 3, which is much less because it doesn't have any faraday cages, is you protect power in your antenna couplings, and the like. Level 4 is basically almost the Wal-Mart- or Home Depot-type of solution where you go out and you buy one nanosecond rated surge arrestors, put those in your facilities connecting to all your equipment as well as things like ferrites that you clip onto your Ethernet and power cables.

So, the mitigation efforts, we have put out guidelines. We're working with FEMA to help cities understand the source region EMP threat. The cities of Chicago, Houston and Baltimore, for example, we've conducted seminars with them. We are deploying and developing infrastructures that are EMP hardened. One of the EAS primary anti-point radio stations that I mentioned earlier currently there's 34 of those stations, are EMP protected. Not in line, like we're eventually wanting to go to, but a station engineer can come back after the event and, for long-term public alerting, we can address that need.

And then we have other programs like the shares program which I oversee and we have Ross Merlin in the audience here who can address that. But we are trying to EMP-harden HF satellite and other communications infrastructures through that. And we do have a full set of models for the traditional EMP threats, whether it be HEMP or SREMP. And I don't have time to go into those right now, maybe in the question period.

Backup slides, I do want to just show that this is not a national planning scenario, but this is what we're using internally, like within the continuity of communications community. This is like in one case they said we need some scenarios to work around. This is an unclassified slide with a one megaton burst over the United States. And it shows damage rings where the orange is bad, yellow is pretty bad, and green is still actually not so good. [laughter] But it's all bad, I hate to say it that way.

And the other thing is, I wanted to show a couple of other slides where you have a multiple EMP engagement. Because you hear a lot in the literature about launches either from subs or ships off the coast, or satellites coming from the southern pole, whether it be launched orbits or whatever, you may have a multi-dimensioned attack that isn't just a single burst going off over the country. And so we have some here like this one is a 300 kilometer burst with a 25 kt, a 25 kiloton. And it's not an enhanced warhead, so most of it is green and yellow. And the next one we have, a one megaton, at the same general location. And you see a lot more yellow and red. And it's covering most of the United States and that occurs if you had a coastal launch, it would occur within about maybe a four-minute period depending on location of assets.

If you do it at a lower altitude, like in this case the EMP commission often would do things at the 70 kilometer level, if you have it burst at 70 clicks with 100 kt, you'll see that most of the east coast is impacted. And that would occur within about a two minute period. So you don't have a whole lot of time to react to these things.

Well, we've to a lot more but I'm going to stop because I've got to give the other guys some time. So thank you for your attention. Next up is Dr. George Baker, professor emeritus at James Madison University. And he's held rather senior positions in government. He used to oversee the EMP protection programs within the DoD. So with that, I'll quickly turn it over to George.

[applause]

Dr. GEORGE BAKER: So, I wanted Kevin to—I'm really glad that he was able to take the most of the time today. He was on the agenda last year and wasn't able to be here last year, so he's making up for lost time today.

I just had a couple of charts. I just wanted to talk a little bit about the source region threat because it's one that's been neglected. It's one of the effects that's been misunderstood at very, very high levels. So there's a prevalent misconception. I see this over and over again when I read some of the scenarios that have come out of the Department of Defense and Department of Energy that "the reason EMP effects are only important if the targeted systems are expected to survive," this is a direct quote, "the primary damage causing mechanisms of blast shock and thermal pulse." And these are what we normally call the hard kill effects and they usually extend only two, three, or four kilometers away from the burst, depending upon the yield.

And then another similar quote, these are direct quotes I've taken out of some fairly prominent government guidance and "source reach is only of concern for military systems that are intended to survive." Both of these are wrong, so draw a red line through those. Those are not correct.

So, when you have a burst on the ground, there is—close in, you have a fire ball, but you also have—just like with a high-altitude burst, you have gamma rays that strip electrons off the very dense atmosphere. The gamma rays don't go very far at sea level, so the source region is only maybe two-to-four-kilometers in diameter, or sorry, radius. So, the source region itself is very limited, very confined in its physical extent. But any long line that runs through the source region will experience—will couple extremely large peak currents running into the hundreds of kill amps close in. And those currents will propagate considerable distances.

I went back through my archives and this was an analysis we did for the Minute Man silos, but it shows the predicted current for an attack where you're seeing it four kilometers out, 25,000 amps and even at 10 kilometers, you've got 12,000 amps. The other point to make about these source region currents are that the duration is milliseconds, not microseconds. So they will cook things.

And then I had just brought along—this is my final chart, that here's a 10 kiloton low-yield burst at 20 kilometers down at a transmission line where Kevin talked about a transmission line running out from a scenario in Washington DC. And you can see that even at 20 kilometers, you're getting a 2,000-amp peak duration of several milliseconds. These are things that we need to factor into our planning scenarios for ground bursts as well. That's all I had.

[applause]

MODERATOR KEVIN BRIGGS: Okay, Mike Caruso with ETS Lindgren. And I'll just turn it right over to you now.

MIKE CARUSO: Okay, thank you. Thanks for having me here. Today, just going to talk briefly about the facility part of protecting the communication system because we've talked about protecting the grid, and that's certainly important. There's a number of things that have to be done all the way down the line to protect the communications capability and the facilities that actually house those communications. The prominent specification that is out there is the MIL-Standard-188-125. However, perhaps, outdated it might be in some aspects, it's what we've got right now and it deals with building a six-sided shielded enclosure around the communications facilities.

What we're seeing very, very quickly evolving is, industry and commercial establishments looking at protecting their operations and their resiliency, their continuing operations. As part of the communications facilities, basically the communications facilities, the old central offices, as they used to exist, are now data centers. And what we're looking at is basically protecting these data centers and most of them are going to try to operate at a Tier 4 level; meaning a 2N scenario where everything is backed up. You've got dual power lines coming in, dual lines going out for communications, dual capability on backup generators.

One of the very important things that I did want to mention is we've talked about harmonics, and harmonics are something that cannot be protected with the standard EMP protection. So, GMD protection and EMP protection are just two separate issues that have to deal with. At this point, we advise our clients to use the very best UPS power conditioning systems they can possibly have to try to deal with the second order harmonics that might be induced by the solar activity.

The filters that are put in place with the protection devices will take care of the E1, E2 very nicely. When building a shielded facility like that, you really have to pay close attention to all the points-of-entry. The slightest hole in the shield can produce the—make the shield absolutely ineffective, and as was talked about, it is not possible to shield the transformer, can't be done. So, I'm absolutely in agreement with that. Anybody that tells you they can shield your transformer, well, move on.

One of the challenges that we have in constructing these communications facilities is antenna interface. And many times, the antenna, or even a microwave length might be in band so basically what our instructions are to our clients is that you really have to have a backup ready, something ready to put in place in case an event does occur. If something is in band, it's just—we don't know of any way of stopping it at this point. So having that at the ready.

We, as a company, just recently completed an 84,000 square foot Tier 4 facility as a 6.5 megawatt facility, totally protected. It's been tested and certified by Little Mountain test facility out of Hill Air Force Base, the pulse current injection. All the components were certified prior to installation, and the installation itself has now been pulse current injected testing.

So it is possible to protect within the limits and within the prescribed areas of MIL- Standard-188 125. As was mentioned, IEC and Dr. Rudasky are working on a more commercialized specification, something that is perhaps more user friendly for commercial industrial world to help protect the facility aspect.

So again, just close in saying it's a multi-step process that we have to look at. We certainly have to look at protecting the power grid, protecting with any means possible the extremely high-voltage (EHV) transformers and the transformers down the line. But you also have to address the facility issues to make sure the equipment within the facility stays running and all of the support equipment can't lose sight of the fact that if you've got a data center, does you absolutely no good to have power up if your air conditioning system has gone down. So thank you.

[applause]

(Transition to Announcement, not on video—) **CHUCK MANTO:** So to catch up on time, Mary Lasky is going to make an announcement in about 30 seconds. You guys are going to be here through lunch at least?

__: Sure.

__: Okay, so we want to make certain you engage them, ask plenty of questions and I'm sure there'll be opportunity for more of that. Mary, why don't you just give us a little opportunity to discover engagement opportunity?

MARY LASKY: Okay. I hope that all of you have found this all very interesting and would like to get involved because getting involved in helping us would just take the pebble that goes into the water, the circle gets wider and wider as all of you get involved. We have some working groups, and in the back of the room there's a list of them, but there's EMP technology, space weather, civilian military, communication technology, healthcare, medical, cyber, education, and legislative and policy issues. You can start your own group. There are cards back there, there's tape. She's standing back there with that.

So with the cards, write your name and email address. Let us know which group you would like to be part of so that we can let the leaders know. If you want to start a new group, tell us that. And so look at the other cards there because maybe you would like to be part of that group, too. So get involved and help us, thank you. [applause]

END OF SESSION

Targeted, Prudent Investments Against EMP: Building Practical Resilience Strategies

https://youtu.be/h5zXsWO6mfE

Presented at the Dupont Summit on Friday, December 5, 2014

Keynote Presentation by Dr. Paul Stockton,
former Assistant Secretary of Defense for Homeland Defense

This keynote presentation describes prudent and cost-effective mitigation measures that the private sector could take using general examples of early adoption of some of these measures prior to establishment of formal regulations and standards. Dr. Paul Stockton, former Assistant Secretary of Defense for Homeland Defense.

Dr. PAUL STOCKTON: Good afternoon, everybody. I'm going to do whatever I can to get us back on track. So I'll keep my remarks brief. But let me start, Chuck, by thanking you, everybody associated with InfraGard, for enabling us to make the progress that otherwise wouldn't be happening. So thank you, again, thanks to Congressman Bartlett, everybody else who has pioneered this sensible strategic treatment of this issue.

What I'd like to do today is offer a strategic vision to prioritize the investment of scarce resources in protecting the power grid against E1, and not only the power grid itself, but the partners who will support the accelerated restoration of power in the aftermath of an E1 event.

My starting point, as always, is the Congressional EMP Commission, which made such a long-lasting contribution. And one of the key findings of the Commission is that when, not if, an EMP event occurs, complex computer-intensive control systems, including those associated with the power grid, they're going to be disrupted. But the vast majority of electrical and electronic systems are going to survive.

The implication here is, we don't need to protect everything. Doing so would be unaffordable anyway. What we need is a strategic vision, a roadmap of where scarce resources can be invested, and then work with public utility commissioners in order to make sure that, from a regulatory perspective, these investments are considered prudent, cost-effective, and therefore, appropriate for cost recovery.

I want to lay out my vision, strategic vision of how we can prioritize investment. First of all, by talking about secure enclaves. I believe that what we're going to need to be able to do is that to ensure that, through hardening and other protective measures against E1, there are going to be enclaves of surviving power, either that survive throughout the EMP event, or can be quickly restarted in order to reenergize the grid, we're going to need to be able to make sure that these strategic enclaves can provide the platform from which power restoration will go forward.

First, let me talk about the two main paths for power restoration that will need to follow in the aftermath of an EMP attack. First, of course, is black start. Utilities typically have very effective, well-exercised black start plans, so that they have generating capacity that, no fooling, will be available to begin reenergizing the grid. And then they'll have cranking paths that gradually allow additional generating capacity to be brought online.

Folks, the targeted investment of EMP protections in these black start generators, and in the cranking path associated and necessary for reenergizing grid, that needs to be a priority for targeted investment. But secondly, as we all know, in previous large scale power outages, 2003, the derecho storms, we don't rely on black start in order to reenergize the grid. What happens is, power is brought from outside the affected area to gradually reenergize the grid. This also is going to be absolutely vital for EMP event restoration.

We need to think about, okay, what are the T-to-T transformers that need to be protected? What are the assets that need to be prioritized for hardening investment in order to make sure that, regardless of which region of the United States is most severely affected by an EMP pulse, by that E1 pulse, that we have a plan both to reenergize from within, using black start and protected cranking paths, and then, from the outside—so again, bring power in from areas not affected by E1 in order to begin from the outside—in. We need both strategies. That's the basis for a strategy, for a prioritized investment of scarce resources, to make sure that we can restart the power grid.

There's one other thing, though, I'd like to talk about briefly. And that is, we all know that, in most severe events, utility companies can't reenergize the grid on their own. They depend on partners. They depend, for example, in Super Storm Sandy, on the National Guard to clean debris, to remove debris, road clearance, security, everything else that's going to be required, these kinds of support missions, by partners, doing what helps utilities restore power more quickly. This is going to be absolutely vital in the aftermath of an EMP event as well.

So we need to not only think, okay, what are the most important support missions that partners can play, that are actually of value to utilities in the aftermath of an EMP event? But this is what I want to leave you with. How can we ensure that these partners, these potential assets, ranging from the Red Cross and other NGOs, to NORTHCOM and the National Guard Defensive Port to Civil Authorities, how can we ensure that these partners, themselves, can survive an EMP attack, so that the vital capabilities they can bring to bear, for example, the case of the Red Cross ESF 6, Emergency Support Function 6, Mass Care, how can we ensure that their vital capabilities are also going to be there when we most need them, when, not if, an EMP attack occurs?

Thank you very much. There is no time for questions. But you all can find me if you need to via Chuck.

Role of EMS and Tabletop Exercises—
Planning for EMP and High-Impact Disasters

https://www.youtube.com/watch?v=qCyccToWGRQ&feature=youtu.be

Presented at the Dupont Summit on Friday, December 5, 2014

Dr. Paul Stockton, Panel Moderator, former Assistant
Secretary of Defense
Dr. Richard Andres, Professor, National War College
Mr. Thomas MacLellan, National Governors Association
MG (ret'd) Robert Newman, U.S. Air Force retired and former
Adjutant General of Virginia
Ms. Cynthia Ayers, former NSA Visiting Professor to the U.S.
Army War College (2003–2011)
Mr. David Hunt, Workshop and Exercise Lead Facilitator
Mr. Dennis Schrader, President, DRS International

This panel reviews the planning and training of federal, state, and local government entities for high-impact disasters, particularly how planning has not been done for high-impact events for civilian critical infrastructure. Dr. Paul Stockton, Panel Moderator, former Assistant Secretary of Defense Dr. Richard Andres, Professor, National War College Mr. Thomas MacLellan, National Governors Association MG (ret'd) Robert Newman, U.S. Air Force retired and former Adjutant General of Virginia Ms. Cynthia Ayers, former NSA Visiting Professor to the U.S. Army War College (2003–2011) Mr. David Hunt, Workshop and Exercise Lead Facilitator.

(TRANSITION TO PANEL LED BY Dr. STOCKTON)

Dr. PAUL STOCKTON: What I want to do now is save the scarce time that's available for our terrific, our distinguished panel. And General, I'd like to start off with you, since you're first. We'll go just in that order. And Thomas, that's not any indication of your relative quality her. It's just the way people chose to sit down. Thanks. Thanks Bob, go ahead.

GENERAL ROBERT NEWMAN: Thank you, Mr. Secretary. Good afternoon, ladies and gentlemen. My remarks are going to be pretty brief, because the Secretary has hit on the two points that I wanted to make. First and foremost, the National Guard is a valuable resource in responding to any emergency. I talk about the typical emergencies being floods, forest fires, hurricanes if you live in Virginia as we do, and the ravages that that can bring to those affected areas.

But the response to an EMP attack is something that is a super response, and one which I have continued to discuss with leaders at the Guard Bureau and within the states with whom I visit, that I think the Guard is falling short on preparation. The Secretary just mentioned support actions, well the clearing of highways, the delivery of fuel, the security around key installations, all of these things the Guard does very well.

But we need to look past the initial emergency actions that are going to occur and look to kind of peel the onion back and discover the second point that I want to make. And that is that we need to be prepared to operate in an environment where the typical support we give is not available. Communications, could they be affected? The ability to talk to units and to direct their operations in support of power or other facilities could be dramatically limited, in some instances, totally eliminated. How do we communicate in those instances?

Another important fact is fuel. We take for granted that we all have diesel generators and those will work. Well, assuming they survive the initial EMP event, how long until those generators run out of fuel, simply because we can't pump fuel to get that into the vehicles? So these are just a couple common thoughts that we've shared among the various adjutant generals that I've spoken with, and had conversations at the National Guard Bureau.

But I think it's safe to say that the Guard needs to go deeper in this, in order to ensure that its ability to support the recovery from an EMP, whether it be to the power companies to get the power back on, or simply to support our citizens that are in need of help, we need to do a better job of preparing that. Thank you, sir.

Dr. STOCKTON: Thank you, Bob.

CYNTHIA AYERS: I guess I should follow on, then, with General Newman's comments.

Dr. STOCKTON: Please introduce yourself.

CYNTHIA AYERS: Oh, I apologize. I'm Cynthia Ayers. I used to be the NSA visiting professor to the U.S. Army War College. But I should preface my remarks with the fact that I'm not speaking for the U.S. Army War College, nor for any other government organization in which I've ever worked.

Having said that, when I joined the military, I was a service member for seven years, not anywhere near as long as the General. And in that capacity, I understood what an EMP was, vaguely. It was mostly rumor. As a Morse operator, I understood what sunspots were, vaguely, but didn't really understand what could happen as a result of sunspots.

I've since learned a lot more about those particular items. And I've also, as of 2005, excuse me, when Ahmadinejad was elected—elected, if you will, to the Presidency of Iran. I did extensive research on him and on what he was saying, and soon discovered that he was talking about an EMP attack against the United States, as were other leaders. But, in this particular case, I was much more interested in Ahmadinejad. He was much more threatening at the time.

So I did a couple of papers for the Director of National Intelligence (DNI), for DNI publication. The idea of an EMP attack didn't seem to resonate with much of anybody. So in 2009, I found myself in Buffalo. And many of you who are out there in this audience right now were giving me information up on the podium, explaining to me what all of this was about. And I admit, even though I knew what an EMP was, and I knew pretty much what it would do, I was horrified. I was absolutely dumbstruck with the possibility of the chaos, the loss of life, the fires, everything that could come from an EMP attack or a solar storm—or a cyber attack, for that matter.

So I went back to the Army War College, and I said, "Don't we need to be looking at this? Isn't this something we need to be doing something about?" And got a lukewarm response, until finally we got some end of year funds. And, in 2010, we put on a small conference called "In the Dark." It was almost the first real public/private conference that we had done. It was just kind of putting out feelers. And it really gave Impact America visibility in that particular regard with the Army War College. But we had a bit more a number of obstacles to overcome after that. And we've been doing so. We have had several workshops and different kinds of things ever since. But they have been smaller. It's been more difficult to get the word out.

This really—going back to the Buffalo Conference, what I immediately knew after hearing what you guys had to tell me, was what would happen immediately upon people recognizing that this was a much more dire consequence than just having the electricity go off for a little bit of time. And that would be that people would start gravitating toward the military facilities that were closest to them, or the police facilities, or the fire facilities. So the first responders and the National Guard, the military in general, were going to be the first people to really bear the brunt of the chaos of what could happen after an EMP.

So got in touch with Dr. Pry and we started looking at that issue. And I've since become immersed. But we really still need to do much, much more in that regard. This is still an issue that will be devastating for all of us, including the military, including the police, and including the emergency services. We need to war game. We need to exercise. We are not going to be able to know the enemy unless we learn about the enemy, whether that be a group of people or an attack mode, unless we study the enemy, unless we inform the people, the population of the enemy. And, if I can leave you with anything at all, it would be that. We have got to inform the public. And they have to know what we're up against, so that they can prepare. Thank you.

Dr. RICHARD ANDRES: My name is Richard Andres. And I'm a professor at the National War College. And I work at a think tank over there. So, what I want to talk about a little bit is what we're doing in terms of exercises. I want to talk about sort of the state of where I see war-gaming exercises on this topic.

So, about six years ago, my boss and friend, who was, at the time, the Secretary of the Air Force, he gave me a Chair over at the National Defense University. And he said, "Take some of these ideas. And I want you to take them out of theory, and I want you to help inject these things into the department. I want you to help inject these into the National Defense."

So I went over there, and I helped, and hopefully over the years supported Congressman Bartlett and some of the other folks that were the leading edge on this issue. And what we did is, over a period of years, we put on a large number of tabletops, exercises, seminars. We invited folks in from all across the bureaucracy. We tried to get senior folks in, as well as real experts from the Department of Homeland Security, Defense, Energy, and from the private sector. And we pulled people together. We tried to be a coordinator to get these ideas and get people talking, get them thinking, and get the people who actually controlled the infrastructure in contact with the bureaucratic entities that regulated it, administered it, and formed the laws of government. So that was our plan.

And I think that we and the other folks that have been working along these lines, that coordinated us, have done a pretty good job. Two weeks ago, the Director of the National Security Agency, Admiral Rodgers, came out and he said, "It's not a matter of 'if,' it's a matter of 'when.'" So these ideas that seemed like science fiction not so long ago, but made it to the point where our national leaders in the United States understand and are working to implement them.

And I'm not going to talk about what the threats are, because we talked about that so many other times. And I think that most of the folks who came here today understand how important these threats are. But, what I want to talk about is the next step. We have leaders of the country, the leaders of the National Security bureaucracy now who understand this. But that's not enough. This has got to get out there. It's got to get into the private sector. It's got to get out into the country. And people have got to understand their stake in this and their ownership in this.

Which means the next level of exercises, tabletops, and war games need to be done by private industry. There are unbelievable sums of money at stake for the banks, for telecommunications, for IT, for many different organizations out there, there's a stake for the insurance companies. And what we need to do now is translate this concern, this understanding about what the risk is, and what the threat is. We need to get specific organizations to understand what their piece of that is.

And that means the next phase in war gaming and tabletops and exercises needs to go out to private industry. We need to get risk assessment from bankers, from insurance companies. We need them to put on exercises where they can bring in stakeholders who can look at this risk that's been assessed, now, by the government, by the military, by our agencies, and translate that into what that means in terms of premiums, what that means in terms of what they're going to do.

Because until this, until people understand the financial stakes that are involved in this, it's not going to move forward. Congress is not going to be able to move forward until folks out there, citizens, understand what their piece of this is, and can then tell those congressmen that they think this is important.

So the next step here has to be moving out to industry. And I think that there are a number of folks here at this conference today who are able to do that, and who are thinking about how to do that. Dave Hunt is one of those folks. So that's a next step. And with that, Dave, I will turn over to you.

DAVE HUNT: Thank you very much. I had the pleasure of working on the Exercise Planning Committee and facilitating the exercise yesterday. I actually wrote all of Rich's comments so I could follow up on exactly what he was saying. [laughter] Hopefully you guys that were there yesterday, which included many—How many were at the exercise yesterday? Just a show of hands. Yeah, we had about two thirds of you.

Yesterday, those of you who were there, you have an idea of my sense of humor. So I apologize, Rich, for making you the butt of my joke. Yesterday, we tried to develop a prototype to follow on exactly what Rich was saying. For those of you who've been here for a number of years, and have been probably frustrated as I am at the glacial pace of this process in educating the public and in raising this issue, there were a lot of people that said, "Listen. We'd like to do an exercise. We'd like to put together a training seminar. We're not really sure—We don't have the technical expertise ourselves to do it. How can we do that?"

So because I'm a friend of Chuck's, we always say yes to anything that's asked. So we ended up coming up with this concept of trying to design a prototype to be available through the EMP SIG to any of the organizations that wish to take this another step further in their own community. And yesterday, we tried to essentially launch that program with the prototype. And we used the opportunity yesterday to kind of have you drink from a fire hose in the entire process. We don't recommend anybody do that. It's too brutal to the participants.

But we wanted to show you what we had developed for your use. And it's a combination of the workshops, educational workshops we had in the morning. It's a combination of the three different exercise tabletop exercise scenarios in terms of cyber, solar and EMP that are available for your use. And again, our recommendation is, don't try to do all three. They're too repetitive. It really doesn't matter why 150 million people are out of power. You still got the same problems. So, if you wrestle in one scenario, you wrestle with the other.

We're also going to be making available a planning guide that follows up on this. So you can take the planning guide, take the materials that we have, and you can figure out ways that you can modify this material to your own needs, to personalize it to your own jurisdiction, put in the name of your energy providers, your local nuclear plants, and whether it's the Bureau of Fire or the Fire Bureau, you can make sure that these things are accurate and relevant to your own needs.

So yesterday, we got a lot of good feedback on what we had done. And I think that we have made a significant stride in the education piece. I've been in exercises for a long time, and I firmly believe that they are an outstanding way to engage people. One of the biggest complaints we had yesterday is we didn't have enough time in the discussions to further discuss what we really would have liked to have done. And that's why we're saying, make sure you don't try to do it all. Leave time for those discussions. That's where you really get the input. That's where you get buy-in from the participants.

Like Rich was saying, I think there's a need to educate the public. We can't do it alone. We need industry. We need our responder agencies to engage the public, to engage our private nonprofit partners. So this will provide an opportunity for you to take this material and move it forward.

The other thing that was available to you as part of the package yesterday was the bibliography of resource material, and also the Read-Ahead package, which provides an overview for any of the exercise participants, particularly those from the private sector and the public who have not been exposed to this material before. It gives them some background information so they can come to the exercise better prepared to participate.

So, in a nutshell, that's what we tried to do yesterday. We thank all of those who participated. We thank you for your input. We will be taking all of that and incorporating it. One suggestion I think was very well done was to take the modules and break them out, so each module you would then evaluate what happened at the end of the first week, what happened at the end of the first month, and then what's the long range recovery picture look like. And that gives you logical breaks to have different discussions on it. And you can break it up so you're not reading 60 slides at once or whatever it happens to be. So I think all those things are very positive. And I thank you for your input and your participation.

THOMAS MACLELLAN: So I'm Thomas MacLellan. I'm with the National Governors Association. I head up Homeland Security and Public Safety on the Center for Best Practices. And I want to commend you, David, that the exercise, the setup with the scenarios was really, really quite excellent. And so I really appreciated it. I actually learned a lot, just kind of thinking about the long-term downstream consequences.

So, at NGA right now, we are just beginning to take a look at this. Not just, we've been on it for maybe a year and a half or so, just as an organization. And again, like the good professor down there, I don't represent any governors. I'm speaking just kind of from my seat at the Division Chair. So I'm not representing any particular state.

But for us, I think the notion that we're looking at is the response. If it's an EMP or a cyber or solar, weather, or something else, a tree branch falling down in Washington State down the whole left coast, we don't really care, necessarily, about what the cause is. I mean to some extent we do. What we're beginning to look at is what the response is and what the response taxonomy for governors will be in the case of a major power outage.

And how that power outage occurs, probably now, for where we are, as an organization, providing governors tools and resources, is probably less important. That said, you have to look at the construct of our government, that when something happens in the country, it doesn't happen in the country, it happens in a state. Locally, too. But in the end, when something big and hairy happens, it really will be the governors who will be responsible for the response.

The federal government, through FEMA, will support the sovereign rights of states. But they do that in support of the governors. And so, you know, the issues that we're talking about here are very unique insomuch that there are big disasters. You know, there's been the hurricanes, and there's been tornadoes. But nothing that we've seen really kind of rises to this level.

And so there are some questions that remain unanswered. What happens—Will the NRF, will the National Response Framework fail under the test of such a major disaster if the whole right half of the country goes down or the middle half or something like that, not even talking the whole country? What happens, will the support functions hold? Will the emergency—will the Interstate Emergency Compact hold? I don't know.

And so, what we're beginning to do, is help governors begin to think through how you will need to respond to these issues when things go down. And one of the things I learned yesterday, and I think I've thought about it kind of intellectually, but it didn't come out to me until about—you know, we're talking about, I think, about the third or fourth week. You've got corrections. You've got prisons and jails. About 1.2 million people, not all of whom are nice, who may be coming out, you know, or getting out, or forcing their way out.

And, when you start talking about civil unrest and government breakdown and so forth, that it really kind of raises some—just that one particular piece, when you look at some of the lessons from Katrina, with the prisoners on the bridge, and you look at some of the sex offenders that they lost, and I believe they'd probably locate them all now. But I know that there was a time when there were folks out that they could not identify, they could not locate. And that raises some pretty serious public safety concerns.

And so, we're looking at ways that we can begin to educate governors about the potential causes, some potential steps they can take to prevent and mitigate. But really, we're focusing right now on the response piece. And I think, you know, some of the Constitutional questions about martial law, and one of the questions that came up in the thing yesterday—I'm not advocating for martial law, anybody here—that you know, can the government take someone's stockpile? Let's say someone is really well prepared, and they have a warehouse full of—their own private warehouse full of stuff that may be essential to the common good. Can the government take it? Maybe. I don't know.

But some of those questions are untested. And so we want to begin to look at ways and strategies that governors can begin to think about how you respond in the event of a disaster. And I want to go back to something that Dr. Stockton was talking about, was the notion of kind of, of prioritizations.

We hosted a work group a couple—not quite a year ago. We brought together folks from the States and private sector and so forth. And we had some discussions, just to kind of educate us and others about some of the issues and what governors should be thinking about it. And one of the key watchwords that came out of that whole discussion was the notion of prioritization, and that when things go down, when the skies go dark, and you lose your power for three months, there are going to be some very tough choices that that need to be made between hospitals or chemical plants or nuclear plants or communication or food production or something. But there will be winners and there will be losers. And we're in untested waters.

So Paul, I'll turn it back to you.

Dr. STOCKTON: Thanks. Well the remarks were admirably short and to the point. I want to thank you for that, because that gives us plenty of time for questions and comments from the floor. All of you are leaders in this realm. We're not going to make progress without you. So, if I could ask that you please identify yourselves and direct your question either to the panel as a whole, or to an individual.

Before we do that, though, let me return to my basic point. I gave you my vision of a strategy for what might work for the prioritized investment in hardening and protection against EMP. Remember, it's going to have to work for industry. And it's going to have to work for public utility commissioners that are responsible for their rate payers, to only allow projects to go forward that are prudent and cost-effective. So, beginning to get down to that level of granularity, to treat EMP, scary as it is, as another risk to manage, with targeted investments that are cost-effective, that's the way forward. Thanks. Who'd like to go first?

BILL HARRIS: I'm Bill Harris, Bill Harris with the Foundation for Resilient Societies. So I have a question for the panel of Chairmen. A brief background, I do filings for this foundation on physical security and cyber security and solar, weather, etc. And when I find that's striking, and we've raised this with the OSD (Office of the Secretary of Defense) officials last week, nobody appears before the Federal Energy Regulatory Commission who represents the Department of Defense, not the General Counsel, not anybody else.

And we get told by the FERC staff, "We need you to file. Because if we don't put evidence in a docket record, the Commissioners are not allowed to use evidence from outside the record." So we've asked them, "Can't you have your staff file?" Well some regulatory bodies do that, but FERC doesn't do that. So we've suggested, if the Department of Defense has a strategy or develops one or the National Security Council, it's not enough to have a strategy. They actually have to send people to the people who set the liability standards, who raise issues, should there be priorities for service to critical infrastructure or DoD missions or specific bases? And people just don't show up. So I think it's a problem. I don't know what you see for a solution.

Dr. STOCKTON: I'll take a whack at that. Let me emphasize, of course, that I'm no longer in office, so that my successors have a chance to take on this opportunity. I think you've raised a very interesting opportunity for progress. Let me only say that, although I'm ignorant of the rules of FERC, I made a point, when I was in office, to talk with public utility commissioners, because we need to, as you know, not only focus on the bulk power system, which is regulated by FERC, but from a National Security perspective, the distribution systems are also vitally important.

So I think there are a lot of partners in the Regulatory realm that we need to collaborate with, to bring along. And absolutely, national security is an important priority to take on. And it may be true that, now I'm speaking as a public citizen, the Department of Defense has done an excellent job of making sure that its own critical components for mission assurance and continuity are well protected against EMP. But installations depend and are part of broader communities. So your points raise very important opportunities for progress.

CAPTAIN DOCTOR KAMANSKI: It's more of a comment on kind of the end result of all of you. I want to thank all of you for your training me on a team leader and Team Rubicon. And I've been deployed seven times, to Sandy, just got back from near Canada in the Washington wildfires. And I helped deploy DMAT (disaster medical assistance team) teams.

I want to just be short and thank all of you for your training at the FBI level, the government level. I just got back visiting with the Surgeon General at the AMSUS [Convention here. (*Organized in 1891 and chartered by Congress in 1903, AMSUS is a nongovernment funded organization for federal and international health professionals. See: http://amsusmeetings.org/about/.*) That's why I'm wearing all of this. I just flew in kind of. And I want to just thank you. That's all. The system is working. And we do support the National Guard and everybody else. And it's because of you I do my job. Thank you.

Dr. STOCKTON: And what's your name?

CAPTAIN DOCTOR KAMANSKI: Captain Doctor Kamanski. I was a former Army surgeon and dentist.

Dr. STOCKTON: Well I'm not sure that everybody knows about the terrific role that Team Rubicon plays. I'm familiar with it, of course. But I want to thank you and your colleagues for their service, bringing terrific military skill sets in support of disaster response, not only in the United States, but nationwide. It's a terrific resource. And would you mind saying just a couple more words about Team Rubicon please?

CAPTAIN DOCTOR KAMANSKI: I entered it when it was fairly small. And you know, what it is, is that it's the gap, that Rubicon is the gap between the immediate response of the first responders, police, fire, and then the long-term recovery. We're there for a couple of months in that gap. And that's about what we do. But the training, you know, my training forensics with the FBI, the training that I received from Andrews Air Force Base at this convention, the system is working. It really is. Thank you.

STEVE CHRIS: Steve Chris at Homeland Security. To the panel as a whole, would you advocate Congress give an additional authority to the FERC to better control, let's say, the NERC, and ensuring that the grid was protected in a more timely manner, let's say?

Dr. RICHARD ANDRES: I'd be happy to take that. Yeah, this is—this is extremely important. This is extremely important, because right now, NERC is doing what you would expect it to do. NERC is a great organization. But NERC is owned by stakeholders who have an enormous—I mean they have pressure. They've got to make a profit. And making a profit means avoiding regulation when possible. They've got to keep their overhead as low as possible, which means there's an enormous incentive for NERC to push back against anything that we're talking about here today. And I've seen this happen. And if you've ever been up on the Hill listening to testimony, you'll know what I'm talking about.

And this has been a—from my point of view, this has been the single biggest frustration that we've got. So, until FERC has the authority to step in and make those regulations that NERC put out there no longer voluntary but mandatory, we're not going to begin to solve this problem. So, to what you said, I'll foot-stomp that, and tell you that that is absolutely key that we do that.

GENERAL NEWMAN: Let me just follow on, Chuck, with that real quick. When you hear all the comments from our guests that have spoken before, and everyone will acknowledge that EMP, cyber, solar storms are a threat, you look at the National Command Authority of our country, the President and the highest [authorities] in DoD, they've taken steps, Mr. Secretary, to ensure that the NCA is protected and can continue on following an attack.

You look at the comments from Kevin before who talked about the Russians look at as a threat, because they buried their systems to protect it. And then you come back, and you look to the citizenry of our country and the fact that we're out there without anyone, despite all the acknowledged, proven facts about the threats that are there, and our vulnerabilities, and nothing's happened.

So I think it speaks to the failure of our system to address what I think is a National Security issue. And what Richard was saying about the FERC and the NERC's relationship, I think points to a key part of that failure, in that we can't have honest discussion without industry input, that, in this case, is negative. It actually detracts from the solution rather than helps it. From a citizen and non-technical point of view, to me, it's obvious that the system that we have, the relationship that is, between FERC and NERC needs to be changed.

CYNTHIA AYERS: I agree. I also would think that not only do we need to grant FERC more authority, but we also need heavy oversight by Congress. And there are reasons for that. But I firmly believe, in this case, oversight is a definite—is a must.

Dr. STOCKTON: I believe that progress is happening now on a voluntary basis. A growing number of energy companies and electric utilities are beginning, on their own, on a voluntary basis, to provide for targeted investment in resilience against E1, as well as E3. And I think we need to give them some credit and understand that progress can go forward, in this way as well. But again, at some point, it needs to make sense from a business perspective, in terms of being able to recover costs for investments in hardening. And that's part of the challenge, too.

Congressman ROSCOE BARTLETT: Roscoe Bartlett, retired public servant. During the Cold War, I attended a forum where the question was asked, what is the most aggressive thing the Soviets could do short of actually launching their missiles? I was surprised at the answer. The answer was, start building a robust fallout shelter program. I wonder if our actions of hardening our grid could be so conceived, and therefore invite an earlier attack?

Dr. RICHARD ANDRES: Congressman. So my background, I'm a political scientist, and I study deterrence issues, international relations. I am not personally terribly afraid that the Russians are going to attack us with an EMP. And I don't think the Chinese will either. The reason I'm not worried about that is because the effects would be so extreme on our country that we would probably retaliate massively. They know that. And so I'm not terribly worried about the Russians and the Chinese.

I am—I am concerned, though, about the Koreans, North Koreans. I am concerned about the Iranians, because both those governments could easily, in the next few years, find themselves about to disappear. They could be in the position that Saddam Hussein was in. And if you were in that position, and you were an Iranian leader or a North Korean leader, you might say to yourself, "You know, hey, we're going anyway. We have nothing to lose. We might as well do this before we go." And I don't think that that's unrealistic.

And I also think that it's possible that some non-state entities, if they were to get this capability, might use it. So I don't think that hardening the grid is going to provoke anyone. All it's going to do is make us safer. I don't see any downside in that way, to a certain—

Congressman ROSCOE BARTLETT: A technical question, to which I do not know the answer. I am sure that neither Russia nor China is going to directly attack us. But, like the Russian ambassador said at that meeting that I was privileged to be at, they would launch from the ocean. Are we sure that we would know the signature of that weapon, so that we could trace it to its origin? You know, a non-state actor can do this. All you need is a trans steamer, a SCUD launcher that costs you $100,000, and any crude nuclear weapon. That won't shut down the whole United States, but it'll sure as heck take out the mid-Atlantic region. It'll take out every city from Boston to Norfolk and west to Pittsburgh. That's a blow that is many times larger than Katrina. I don't know if it'd be a fatal blow or not to our country. But can't they do this without leaving any signatures, so that we don't know who it came from?

GENERAL NEWMAN: Congressman, I'll tell you that this attribution is a huge effort, issue in cyber. And I know finding out who threw the switch and who caused the problem is a deterrent in itself. When they can mask not only the vehicle, but the culprit themselves, it's a huge advantage for them and emboldens them to take steps. I can't comment technically on whether or not we can firmly establish who launched something like that. I know our cyber efforts have gone a long ways in enabling us to assign attribution to this. But it's still a mystery.

I think it's a priority for us to continue to work on this. And I think, Mr. Secretary, if you know something different, please correct me, but I think cyber command, this is one of the key nodes they're working on, trying to identify where these attacks come from, because the anonymous nature of these attacks makes those that would launch them feel a lot safer in doing so.

CYNTHIA AYERS: I would like to echo Congressman Bartlett's remarks. I think not only is there an attribution problem that we need to be thinking about, but unfortunately, by the time we would be able to establish who it was that attacked us, it would be too late. It would be irrelevant for us. So I do think this is a very big possibility. I also think that it's a possibility that Russia or China could be indirectly looking at this, and coalitions have already been established, that I think we ought to be looking at heavily, coalitions with Iran and North Korea by the Russians and the Chinese. So I agree wholeheartedly with your worries along those lines. Thank you, sir.

Hon. ANDREA BOLAND: Hello. I'm Andrea Boland of—recently a state legislator in Maine. And I wanted to just—who brought this bill to, as some of you know, who brought a bill looking for protections at the state level. A number of states now are trying to take different sorts of actions to also find a way to protect their grids. The main problem is NERC, because they set a profile standards that are totally unhelpful, and some would categorize as criminally negligent. And they know the extent of the threat and continue to downplay it.

And we have seen it up close in Maine for the last year and a half, that even having their own data from the grid in Maine, don't even choose to use it, but instead set what they say should be a lower standard, which initially, and perhaps they'll return to it, they go back and forth. Said to them, "Don't worry. You're fine. We don't need to do anything in Maine."

And all the work that we have done in the State of Maine, and that others are contemplating doing in other states, can be just turned upside down by the Public Utilities Commission in conjunction with the power companies, just say, "Well, we don't agree with what these independent experts who have done stellar work are saying. NERC says this, we'll just go with this, and we'll keep you vulnerable, essentially." Would you contemplate coming up with a different kind of an agency, rather than NERC, to set the standards so we don't really have the fox guarding the henhouse?

GENERAL NEWMAN: I think the answer to this might be the reestablishment of the EMP Commission. Now Cindy was talking about this earlier. But I think we need an unbiased, recognized panel of experts whose opinions can be valued and without prejudice, make recommendations based on the facts. And I'm hoping, perhaps, that the Senator from Wisconsin who spoke this morning, Senator Johnson might lead the charge on this. He seemed willing to hear more information about it. And perhaps it's time to reengage that.

I think we've come a long way in the last five years, in particular, about a recognition of this. Maybe it's time to reengage and have a candid conversation. And then, most importantly, have some action based on the findings of that commission. Cindy, anything you or anybody else—thoughts on that?

Dr. RICHARD ANDRES: I just want to—Andrea, thank you, and thank you for all that you've done on this. I think everyone's aware of all the work you've done and how hard you've pushed and some of the frustrations that you've run into. Look, NERC is a great organization. It's a private organization. It's an organization of private companies doing their best. But they suffer from a collective action problem. And that is, if you're a member of NERC, and somebody says, "Look, you're going to have to raise your rates. You're going to have to take these extra precautions," what happens? You're going to lose out to your competition that doesn't have to do it.

NERC will always suffer from that. NERC is the fox guarding the henhouse, for good reasons, these are good people. But the way that it's set up today, NERC will never be able to solve this problem. And, as long as NERC is the authority, and able to say these regulations are only voluntary for you utility companies, NERC will not be able to solve the problem. And, more than that, NERC will always stand in the way of any solutions that anyone brings up, unless—unless Congress can come together enough to say, "All right, all of you guys, everyone has to meet the same standards. Everyone out there, NERC, you've got all of your members have got to do this." In that case, nobody has an advantage. Everybody is living under the same rules. So they're going to adapt to it. And then they're going to fix the problem, because NERC has the technology, it has the wherewithal, it has the capital. It is industry, right.

But you've got to get Congress to tell them, "You've got to fix the problem. And it's not voluntary, it's mandatory. And here is what you're going to do." That solves the collective action problem that they're dealing with at the moment. But yeah, I absolutely agree with you. As long as NERC is guarding the henhouse, that's going to be a problem.

Dr. STOCKTON: I just see things very differently. Somehow, despite this problem, a growing number of utilities are investing on a voluntary basis and hardening their critical systems against E1. There is no regulatory requirement for them to do so. And yet, a growing number of utilities are doing it. And I think this is another model. It's another path forward. And it's a way to soon, to give up on this kind of collaborative progress. And I would hope that we'd be able to sustain that, as well.

JENNIFER BREZOVIC (managing microphone for remote InfraGard group in Charlotte, NC): I'm going to bring the microphone over here. But I have a question from Charlotte, North Carolina. **STEVE VOLANDT:** And this may be something that, if this panel does not care to answer, perhaps another panel coming up would. What's the plan, and who's responsible for nuke plant waste fuel cooling plants, and chemical plants, and refineries that lose power and may catch fire? And we have another group that might be able to answer that. But if somebody wants to take a shot at it now, they may. And then I'll bring that mic over to my right.

CYNTHIA AYERS: I might suggest that Tom Popik really the best person to answer that, or Bill Harris. So I suggest Tom Popik or Bill Harris actually answer that, in regard to their input.

CHUCK MANTO: They'll have an opportunity in just a moment.

DAVID BARDIN: I'm David Bardin. I'm a retired volunteer. But one piece of my biography is I worked for FERC's predecessor, from 1958 through 1969. I remember when NERC was not the corporation, but the newly created National Electric Reliability Council. I remember that Sunday in 1965, when I happened to be at the office, my wife called up to say, "Something funny is going on. All those places north where we used to live, and our friends used to live and are living, the lights have gone out." That was called the Great Northeast Blackout of 1965.

So, with that in mind, Dr. Stockton, I've got a question for you. I understand your point, abstractly. And I think it's wonderful that some utilities are saying, "We're going to do more, better than the FERC regulations based on the NERC recommendations require us to do." I could even recite, because of a letter I've seen from one utility to its state regulators, one name of one utility.

But it really doesn't do me much good just to say that in the abstract or to stop at your statement. Can you tell us the names, not necessarily of all the great utilities, but some of those utilities that are setting an example of what I would call best practices? If the FERC regulations based on the NERC recommendations aren't really the best practices, who are some of the utilities we should be patting on the back and holding up as examples to the rest of the utility industry?

Dr. STOCKTON: I think that we should take that back, Chuck, for action next year, invite utilities to make their own case, to talk about what are the best practices, or more, what are the smart practices that are being developed by industry, that now make sense, from a business perspective, and broaden this dialogue so that those who are making progress can share some of what they're doing.

Now, some of these technological solutions may be proprietary. And that's why I'm not going to comment on them today. It's for utilities to decide how much to share. But I do think that building more effective sharing mechanisms for emerging technical challenges that make sense from a business perspective, that's got to be part of the solution.

CHUCK MANTO: And we have time for a couple more questions or so. We've got people in sequence. I do know that there are some companies who have actually been active with our work, and some of the people in this room know who they are. In some cases, they say, "Please don't tell anybody yet." But they sort of brag on the sly. So it's not totally unknown. There is some work being done. But I think I had a utility person saying, "Everyone who is a utility wants to be first at being second." And so I think if that's true, then when we get a few of these folks doing best practices, it may compel people to do more. And it certainly still doesn't inhibit anyone from declaring this best practice or something close to it as a new standard, as opposed to let's keep it as benign as possible. But we have another question. Please raise your hand.

JOHN ROSICA: Yes, good afternoon. John from California in the telecommunications private sector, also with InfraGard. Just a quick thing for Congressman Bartlett's question. I asked the exact same question yesterday at one of the experts. And the answer on the signature was that yes, the byproducts of these explosions can be signatured and mapped for proper interdiction of follow-on event, as Kevin said. But my question goes to the response side of the equation. One entity that seems to be lacking, both yesterday and today at this event, and they'll be the first to stick a camera or microphone in front of your faces to get their snapshot when something does happen, is the press.

And my great concern, as has been evidenced over the last few years with the disasters of various types that we've seen in this country, and I'm going to be uniformly insulting to all of them, is that they completely lack subject matter expertise. And they completely lack the ability to keep the public calm. And my question is, to the governors, panel, gentlemen, and anybody else who mentioned on the response side, is there a plan also in parallel being developed, other than what we have in the form of what we call the EAS at the federal level, that in California we've seen used a number of times recently? And it seems to work. But overall, in a prolonged outage, we're going to need competent, credible, calming information. And I don't think—and I hope I'm not going to get shot here—that our current state of the press is going to provide that, by any means, by any stretch.

THOMAS MACLELLAN: So with respect to—I'm not going to comment on the press piece, because that's just out of my wheelhouse. From the governors' side of things, I can tell you the governors take their responsibility as communicator-in-chief very seriously. And we spent a couple weeks ago, we do what we call the Seminar for New Governors. When governors come in to office, the first thing that we work with them on is being prepared, Day One, to respond to a disaster, whatever that disaster is, manmade, terrorist, natural.

And one of the cornerstones is working out those communication strategies so there's a unified message that you're, if you don't know, you say you don't know, but begin to build those things out immediately, from Day One. And we spent a whole day. They got a classified briefing. We were able to do some work with them about thinking through where you sit and where you stand, how you communicate, what words. If you don't know, you don't know. How you access information.

And so, from our perspective, we work very closely, and the governors take their responsibility as communicators-in-chief very, very seriously. As to the press, I can't really comment to that.

DAVE HUNT: [Let me take a whack at that.] One of the things I've done over the past several years is spent about five years working with states, major cities, regions, and the development of public information strategies and coordinated public information messaging. And one of the things that we try to do, and it's very difficult in a large region, is to get a single unified message out in a timely fashion. It sounds very simple, but it's actually very difficult.

One of the advantages we have today is we have the ability to go on the web. We have the ability to use the tweets to get a message out directly from the jurisdiction to the public. We can bypass the media, in that sense. And it's not a negative way. You can also feed the media, but you don't have to rely on them to convey the message and to put a filter on the message.

So it seems to be working quite well in terms of getting the message out. There's a lot of skill, now, over the past several years that's been adapted. We see a huge following on the Twitter feeds that are coming out of the organizations. The other half of your question is, what happens in the event of a long-term power outage and the ability to use this? Well, all that is negated. And the ability, then, to communicate with the media is likely going to be from the mayor or the governor or county commissioner, whoever it happens to be, in a face-to-face thing. And all bets are off right now. So there are good strategies currently that are being in place and improved on a daily basis. But an EMP, all bets are off.

CHUCK MANTO: We have one more web question, and then we'll have to take a stop. So I'm going to reword this. Can the Critical Infrastructure Protection—I assume Act—be extended to cover this issue? Isn't it enforceable? Of course it only passed the House, if that's the Act referring to. But, since we've been talking a lot about the Critical Infrastructure Protection Act, which will compel DHS to study this, and to do R&D in this field, is there a sense from any of you or a couple of you about the effectiveness of this proposed Act that's just been passed by the House and is now hopefully going to the Senate, at least hopefully in the minds of the people who have passed it in the House?

GENERAL NEWMAN: I know that Virginia's legislature will take this bill up. There is a bill pending at the current legislature that will address the same issues that the House has brought forth with the CIPA Act. I think the key will be what we do with it. We can study it. We can exercise with it. But if we make no changes to the procedures we use to address it, I think we'll be back in the same boat we are today.

DAVE HUNT: To echo that, if the bill passes, there will be an additional scenario added to the national list of scenarios that can be addressed by locals and states. There is no requirement for the states and locals to address those. And right now, there is—you know, there is a debate on how you plan. You can either do capability-based planning, where you look at what capability your jurisdiction would need to respond to an event, and then you build that. It means what type of equipment, what type of staffing, training, protocols do you need to handle it.

And another way to evaluate that, which is what I recommend, is then you go through the different scenarios, and you evaluate how your capabilities that you have developed would stack up in the event of various scenarios. There's no requirement for any jurisdiction to do that. There's also no requirement, that I'm aware of, in the bill, for DHS to fund this going any farther than adding the scenario. So, in my opinion, it needs to be more than just the scenario placed in the list of 16 scenarios that they should be able to respond to. It needs to go further than that.

CHUCK MANTO: Dr. Stockton, would you like a concluding remark for your panel this morning? Thank you again for coming.

Dr. STOCKTON: I would. I would. And I think that's a great way of wrapping things up. EMP is just another high-consequence risk to manage. That's treated as such. There are sensible ways to move forward and make progress. Our goal, now, is to accelerate the progress that's already been happening, and to work together to make sure we assume—we have to, from a prudence perspective, that this might happen. That, if it does happen, we'll be ready to be able to much more rapidly accelerate the restoration of power than would otherwise be possible. So thanks to all of you on the panel. And thanks to you for the progress that you're helping make.

[APPLAUSE]

END

Resilient Hospitals in Large-Scale Disasters—The Role of Alternative Technologies and Sustainability in Electric Power Grid Mitigation

https://www.youtube.com/watch?v=HD1mhIKstAc&feature=youtu.be

Presented at the Dupont Summit on Friday, December 5, 2014

Dr. James Terbush, Martin Blanck and Associates, Public Health, Colorado Springs

Dr. Terry Donat, Independent Biosecurity Consultant an IEMA RACES Officer, Chicago

Dr. Donald Donahue, Consultant to American Academy of Disaster Medicine, Washington DC

Ms. Sierra Bainbridge, MASS Design Group, Boston (Off-grid hospitals in Rwanda and Haiti)

Mr. Art Glynn, CAPT USN Navy Emergency Preparedness Liaison Officer, USNORTHCOM

This panel compares the work underway in the U.S. DoD, the private sector, and efforts in third world countries for hospital operation without the benefit of power grids. Dr. Terbush who previously worked in this area for the U.S. Northern Command leads the session. Sierra Bainbridge discussed the role of off-grid hospitals in Rwanda. Captain Art Glynn discusses the DoD SPIDERS microgrid program and Dr. Donat discusses local emergency communications in the Chicago area.

Dr. JAMES TERBUSH: Thanks, Chuck, and thanks also to the organizing committee for the Dupont Summit. Can you hear me okay? I'm usually pretty good with an outside voice. Okay. And in the last two days I have heard a number of comments about health and health-critical infrastructure. In fact, yesterday during the exercise it came up quite frequently. That warms my heart. We would like to be higher on the priority list and know that it is in fact an essential function, an Emergency Support Function #8 that we in health and medical have.

Last year I had the good fortune to be able to present on a subject of public health consequences of a cyber-event, which included EMP and solar storms, and this year they invited us back to explore the topic of health and some of these major threats a little further. The thing I particularly appreciate is that these subject matter experts who are here with me today have a broad grasp of the threat, but also the technology and some of the solutions, and that is sort of the guidance I gave to my panel, which is that we have discussed threat a lot and we have a pretty good grasp of some of the physical consequences of these phenomena, but what I wanted to ask our panel to focus on today also is some solutions, both technological and others. So that is what we're going to do today. We'll hit on some particular threats to health infrastructure and then focus on solutions.

So, the panel, in short, let me go ahead and make some introductions here. First of all, I would like to introduce Dr. Terry Donat, and Terry has done a lot of things. He is based out of Chicago, but he particularly is going to talk today about communications, and we all know in whatever disaster we're responding to communications is number one, two, and three. So, health-critical communications. Dr. Don Donahue I've known for a long time. He has done many things in his career, but he is a consultant to the American Academy of Disaster Medicine and also involved in many private sector, government consulting. Ms. Sierra Bainbridge, Chuck had indicated, we were hoping that she was here. I spoke with her just yesterday. And she may have gotten tied up. There is a lot going on in town right now. So we may yet get to see her. If not Chuck has promised to sort of fill in on their encounters in Africa when he was over there. And Sierra's topic is off-grid hospitals, off-grid hospitals. And then my good friend, Captain, U.S. Navy Captain Art Glynn who has a unique perspective straddling both the private sector in his work with Booz Allen Hamilton and also as a Navy Emergency Preparedness Liaison Officer at U.S. Northern Command.

Quite a Colorado contingent at this conference. Can I see hands of people who are from Colorado? There we go. Well, there are a few. Okay, great. And appreciate what you do there in Denver.

Okay. Well, my comments are going to be brief, and let's just say that with increasing complexity comes increasing vulnerability. Now, some of the solutions to our collective problem may in fact be advanced technology, but in the medical realm it's easy to imagine that over the last 30 years or so that the way that we practice medicine is different and that we are increasingly reliant upon sophisticated technology. And I have the good fortune of teaching medical students, and so I am finding that in many cases we're losing some of our practitioner diagnostic skills again in favor of scanners and other types of technical equipment.

Add to that the electronic medical record. You may remember from the news quite recently that in Dallas the first Ebola patient to die in the United States, there was a little bit of confusion about notes made on an electronic medical record in the emergency room, and old school or old style where we had a paper record that might have not been overlooked. By contrast, in Denver a little over a year ago there was a mass shooting, unfortunately, and the University of Colorado Medical School, not too far from there, received the majority of the patients, and they had at the ready paper charts, gurneys, and of course they had practiced doing things, again, sort of old school.

So, one of the appeals, one of the solutions, one of the possible solutions that I would like to offer in the healthcare sector might not be more technology but less, sort of a return back to the future in that we would then hone a lot of the skills that we learned perhaps when we were earlier in our careers and practice those without benefit of a lot of the sophisticated equipment we use in hospitals.

Recently with Hurricane Sandy, with Super Storm Sandy one of the major hospitals, New York University Hospital, was without electricity for well almost 30 days. They had gotten over the generator in the basement thing, but in turn, some of the switchboxes and other apparatus were flooded when the storm came through. Because there were other hospitals in the area, the patients were transferred out and there was no loss of life that I know of from that particular hospital evacuation, which by the way is a really tough thing to do.

But if it was a widespread disaster we wouldn't have other hospitals in the immediate area with the sort of catcher's mitt on, and hospitals would have to stay and play, if you would. They will have to continue their operations, which again is another theme of this panel.

Again, on the positive side I would like to mention that there is some policy changes in the works as a result of Super Storm Sandy, and specifically with regards increasing disaster preparedness. This particular document, which is a draft, I know is being actively worked on. We talked with one of the authors this week. And it has to do with 17 different categories of healthcare facilities that will be required to make changes to their policies and procedures regards disasters and specifically disasters in the light of lessons learned from Super Storm Sandy.

So, their hook, if you will, is that they're going to tie Medicare and Medicaid payments to these changes. I would suggest also, and this is a good time to do it, that this new legislation, CIPAA, which would include the EMP scenario, hospitals and clinics, and emergency management personnel, of course, trained all the time, that they now additionally train to an EMP scenario. So, this might be something if we're lucky and timely we can get instituted into this rule making.

The only other piece that I would add to our policy discussion is personnel changes. In most business it's the people that matter, it's not the things so much, and particularly in the hospital business we have looked at this, and in a particularly stressful scenario, such as a pandemic, there may be as many as, as much as 60% absenteeism of medical personnel, to include people who stay home to take care of their own families. You may think about your own business sector and may come up with similar numbers, but these are the, these are your first responders in the hospital.

So we need to do things better on a personnel policy basis to encourage people and allow them to be able to come in and do their job and do their job for you, which may include such things as helping to protect their families. So, in an EMP scenario it may mean that healthcare facilities need to also not only make their hospital facility more resilient but allow the family members of staff to be able to come in and get help as well.

So, I think I have—I'm not going to touch on some of the other things here, but I would like to go to our first speaker, Terry, if we could get his presentation up, his slides. Yeah you do. That's one of the three. Dr. Donat, I know, is a very resourceful guy, so he may start his presentation acapella, and if we could get the slides as soon as we can. You want one of the other panel members to go first? Okay, Don, can you? Can you go? Are you Sierra? Welcome, Sierra. We're definitely doing this one on the fly. That's because we're resilient.

Dr. DONAHUE: We're resilient, yes. You see slide one, it says, and I'm going to expand on them some, a lot of stuff that Dr. Turbush just talked about, that increasingly our hospitals are wired. The trend in healthcare right now is to move away from hospital care. Care is moving out of the hospital. It's moving into the community, into the ambulatory care world. But when the push comes to shove, when I'm really sick or something, everything starts breaking down, we go to the hospital.

So if you look at, if you look at the modern hospital, unlike the example in Rwanda, our diagnostics, your x-ray machine, your MRI, your lab all is highly, highly automated. Your climate control is important. It not only keeps the air hot or warm, but as alluded to, it's your infection control mechanism, big issue with Ebola and Middle East respiratory syndrome (MERS) and some of the other communicable diseases that have shown up. All your mechanical, your doors, your access into your pharmacy, your elevators. If you can't get between floors you can't push that gurney up and down to the OR.

And then electronic health record. Not only are electronic health records becoming prominent, as a nation we have legislated that. If you're not using electronic health records in a meaningful use, you will be penalized in the future. So, all of that is pushing us to a highly automated system.

It is a great visual. I called it Lessons Earned. In 2005 hospitals in New Orleans were flooded. There was one hospital in particular, Methodist Memorial, that they were smart, they said, "We'll put the generator on the roof." Out of all the hospitals in those three perishes they got it. Unfortunately, their fuel pod was out on the edge of the property, so they went down and they flooded. And so the visual here is lessons earned. So in 2005 we flooded a hospital and guess what? In 2012, we flooded hospitals in New York City again. As a culture we are really slow to pick up. It's sad.

So, absent hardening. I have a great picture here of Godzilla. Unless we harden, do all the things that you have been hearing about for the last two days, the modern hospital will not function following an EMP, a CME, or even a really big lightning strike, an approximate lightning strike. Our modern hospital will be offline. But the problem isn't necessarily what you think it is. I put the one too many pages.

I had another, an aerial shot of a major medical center in the Northeast that I won't identify, and their power went out. The power went out and all their backup systems, generators kicked in and they were tooling along just fine. Then at the edge of this major campus there was this little concrete block building that just sort of innocuously sat out there. Well, it turns out that the little concrete block building was the pump house for the sewer system. And if you have ever lived in a house with a septic tank and it gets full you know what happens.

Well, the hospital starting on, I think, the fifth floor trickling on down, this major medical center had a huge problem. Nobody planned for that. The OR ran. The ER ran. Nothing would flush. So your problem isn't necessarily the thing you think it's going to be. So the way ahead is, and we have talked about the way ahead is one, and I'm probably preaching to the choir, the health system, you need to do a comprehensive vulnerability and risk assessment, but really comprehensive, looking beyond the normal things. We have a strong tendency to say, "Oh we're in charge. I run disaster drills. We know what is happening." Look at your internal systems. Look at your external systems. Where is the fuel for your generators going to come from?

One of my pet peeves, where is your oxygen going to come from? I'll bet you in every town every one of you is from there is one, maybe two O2 companies. So if every hospital in your state needs an extra load of oxygen where is it coming from? That's an unanswered question. And then start to develop, Dr. Turbush's, you need to start to develop viable alternatives for staffing, for access to the hospital, to the facilities, your outlying clinics, modify standards of care. The standard of care for being on a ventilator is 100% oxygen. What do you do when you're running out of oxygen?

And we train our physicians and our nurses and our medical staff so well to quality of care, when it's time to alter, to work outside of that parameter they have a hard time with it. They really have—even in the face of disaster they have a hard time adjusting. So you have to work on that and you have to train them for the unexpected. And then you have to plan for working around healthcare probably more than anything uses just-in-time supply processes.

In a major blackout there is not going to be any time for the just-in-time. A couple of other slides talking about consider capital investments. I actually had the privilege of sitting in on a briefing for a company that does construction, and one of their main lines is retrofitting and making hospitals, taking them off the grid, which leads into your presentation, taking them off the grid, making them self-generating, self-generating their own heating and air conditioning. And there are actually incentives and there is a business case that that saves the facility money. It is a capital investment that in the long run it saves you money and it gives you resiliency.

Harden your electrical systems. I was having a talk last night with a colleague in Tampa and I said to her, "You know, in your house is your television, your stereo, your microwave, are they all on surge protectors?" And it made me start thinking about my own house. And I said, "You know, there are simple answers that we just fail to follow."

Plan to be able to operate for extended time off the grid. Anticipate your external pinch points. Expect an increased patient load. When things go bad people start doing stupid things. People start running generators inside houses, heating their house with charcoal. Puncture wounds. So when things go bad more people are going to come to the hospital, not fewer. Understand and train for the modified standards of care. Try to be self-sufficient. In all likelihood help isn't coming. You're going to be working on your own. And expect the unexpected. I have a note here that says consider Joplin. Nobody but nobody that I know of predicted a two mile wide tornado that would knock a six story hospital off its foundation. We never planned for that. Plan for the unexpected. It's coming.

I have a slide that it's not a matter of if, you have been hearing that all day, and I had another great slide that I stole from one of the speakers last year. It was a great old—there is a picture here of a stone in Japan and I don't know, very old, and on the stone, in Japanese of course, it says, "High dwellings are the peace and harmony of our descendants. Remember the calamity of the great tsunamis. Do not build any homes below this point." They didn't follow their own advice. We tend not to follow our own advice.

We need to look at the situation and think of the unthinkable.

Dr. TERBUSH: Thank you. Thanks, Don.

[Applause]

Dr. TERBUSH: Okay. And I am very pleased to introduce Sierra Bainbridge. We have spoken over the phone only before, but I was so excited about the product for MASS Design Group out of Boston and what they have been doing in Africa. So, Sierra, I would like to invite you to go ahead and tell us about off-grid hospitals.

SIERRA BAINBRIDGE: Hi. My apologies to the panel for being late. I am so happy to be here with you all, and also hopefully to learn as much as possible. And I apologize as well, my slides also are somewhere in the ether and have not landed, which is unfortunate because of course I am an architect and a lot of the things that I was hoping to show you were images of some of the work that we have done, which is kind of fun to break up the information that is passing.

So, I am going to go ahead and just talk a little bit about our work. As mentioned, we work mostly in Africa, also in Haiti, and we're just starting to work in the United States a bit. And our goal is really to think about how we can take the lessons that we have learned in those places and bring them to bear on the work that we hope to do here. And especially in light of some of the issues that we have been having in terms of major disasters that really cause us to be in peril of lack of care during those times. So, and many of the ones that have just been mentioned I think are the examples also that we're looking at, and especially Hurricane Sandy. Our office is based in Boston and so we are thinking a lot about the storms that are bound to hit in our area in the near future.

So, learning, we think of it a little bit of south-to-north learning, and what are the things that we can take from the work that we have done. Specifically, we're working in a place that has very little guidelines. A lot of the countries that we're working in do not yet have healthcare building standards and codes, and in fact we're working on writing some of those for countries like Liberia. Unfortunately, it only finished in last December, so they were not able to begin implementing that ahead of Ebola. And also the same for Rwanda.

In this case, it actually frees us up to really try to do things that might not fit in, and I think that's one of the limitations we have here in the United States is obviously we have huge amount of code and infrastructure that are to create the safest environment possible, but often that ends up limiting us and creating some paralysis in our ability to maybe do the best thing possible in our current moment. And it takes a little bit of time to kind of dismantle those or loosen those up so that we can provide the right kind of solutions.

We are dealing with a lot of other limitations, very, very low budget on all of our projects. We're looking at very limited resources in Rwanda, for example, importing, there is no local steel or cement availability, so we import a lot of that, which means very high costs for those items. So we're always constantly looking for what can we use locally. And I think, I mean that applies everywhere. But I think most pertinent to this discussion is we are looking at places that don't have a reliable grid, so the electricity continually goes off multiple times a day or consistently at a certain time each day, and/or it will kind of move around the particular city, some places are on and some are off. And part of that is in Rwanda all of their electricity at the moment is brought in on a truck through Diesel fuel to one central location and then it is kind of pumped out to the rest of the country. They are looking at methane gas as an alternative, but until then it is extremely expensive, and so they kind of have to ration it out.

And so the result is that we're looking at ways of building hospitals that can function still when there is no electricity and on the kind of very least level of resources possible, and we're still having a lot of these diagnostic equipment and other things. But to kind of state the most obvious steps that we have taken, starting with Butaro Hospital, which we started in 2009, that's when I moved to Rwanda to work on that project and finished a couple years later in 2011, and this is where I would be showing you the great images of this beautiful hospital. I mean it's really one of the most beautiful places I've ever been. Rwanda is called the land of a thousand hills, and it's extremely green. And this is really sitting atop of one of these hills.

And so we were able to take advantage of some very simple design moves in order to increase the resilience of this hospital. So, one is natural lighting and ventilation. We still worked with all of the guidelines from WHO and we worked closely with doctors from the Harvard School of Public Health who were really focusing on infectious disease control, especially around the areas of TB. And we went back and forth with them throughout the entire design process to really make sure that we were covering our bases and creating a redundant system by which if the electricity is out we still have a way to get minimum 10–12 air changes per hour, so we can really make sure that we are continuing to keep infection control measures going.

So, we took a few main moves. I mean another bit of context. I will wrap up as soon as—I have a couple of other—.

Dr. TERBUSH: You're good at doing word pictures, so go ahead.

SIERRA BAINBRIDGE: Oh good. TB is a huge issue in Rwanda, and that, or not in Rwanda, but in Sub-Saharan African, and TB specifically is an airborne disease and so the air changes per hour really have an impact. And people are in fact going to hospitals, catching things like MDRTB, which is multi-drug resistant TB, and dying from actually going to the hospital without TB, catching it while they're there being treated for something else, and then dying from that. So it's a real threat. And so this was the focus of the design behind this.

We did some very simple things. Because it's the climate was quite temperate or tropical temperate, meaning it's about 70 degrees every day, we moved all of our, all of our pathways, rather than having hallways on the interior of the building where people often wait and that's where diseases are transmitted in the waiting areas, all of those were moved to the exterior, so we have no interior passageways. All patient waiting is also on the exterior and we created a beautiful landscape to still make these places really palatable and really comfortable and very dignified, but reduced that potential for transmission.

In terms of the air changes per hour in each of the rooms we worked really hard to have excellent cross-ventilation for our primary method of moving air through, and we rely on that for our multiple air changes per hour. And then in addition we have, and this should be running all of the time when we have electricity, we have huge two meter wide fans. We have got the hospital. Excellent. Can I—We can skip through quite a bit. This is some of the introduction. Yeah, very good.

So this is Butaro. And keep going just a little bit. I don't know if it can be taken over here or not. A little context.

I don't know where to aim it. So this is the kind of context we're talking about. Again, really sitting on top of a mountain in Rwanda. Here you can see all of the pathways, as I mentioned, have been moved to the exterior of the building, but are still covered, so you can move between every treatment area under cover, but on the exterior. And again creating these large expansive landscape areas for people to wait.

The question is, how can we create these redundant systems. And I'm catching up now. It's still coming in, I think. Let me point it in the right direction. There we go. This way, okay. Perfect. Thank you.

So this is, just as I was speaking about, we have operable windows here, fixed open louvers, which is creating our cross-ventilation. And then most of the time when we do have electricity we have these very large diameter fans that were actually donated by Big Ass Fans, this amazing fan company, and ultraviolet germicidal irradiation (UVGI) fixtures. So, this brings the air up into this zone, this plenum of air, and it is treated there by the UVGI fixtures. However, when it is off we still have 12 air changes per hour. So these are really simple methods of having systems that still allow us to have infection control even in the case when we don't have electricity. And this is just a section showing the same information.

And we have tried to use this same work on—thank you—on our other work that we have been doing in Rwanda and in the area. This is a huge maternal services expansion for a hospital that had a very, a building from about 1910 which was in very poor repair. And we have used these large chimneys here for the same purpose. It's an urban area, so unlike being on top of the hill we couldn't rely completely on natural cross-ventilation on the ground level. And so by lifting this chimney up and using convection, which I have a—I keep pointing the wrong way to flick the slides. Okay, back one. There we go.

We're using this dark, dark stone and this roof that is painted black to really pull the air up and out of the room. So we still are having, we're still managing to get those air changes per hour, even though we are in an urban area and don't necessarily have a breeze at all times. And so these are the kinds of very simple systems that we're trying to implement on each of the hospitals.

This is a large hospital, again in an urban area in Haiti. This is a competition, but it still speaks to the concepts that we're talking about. Specifically, here we were looking at on the lower levels where you have a very wide slab, we would put all the technical services down there, at the top of the building, where you have a much narrower slab. That's where you would have wards. So you can still have cross-ventilation where patients are spending the majority of their time.

And along the back of the building, and in all of our projects we always try to have ramping; even on a building, this one is multi-story, most of ours are one to two stories, so that even in the case of an electrical failure you can still manage to get people from floor-to-floor as needed. So this allows for gurneys and in the case that your elevators are not working you're still able to function as a hospital.

So, there is a third system. And then obviously we're trying to keep on all of these buildings as much light in all of the spaces as possible so we can rely on natural lighting. And then I'll—This is just some diagraming of the same information. And I'll just talk about this one last one and then I'll be finished. This is a cholera treatment center. It's also a diarrheal, now a diarrheal disease treatment center, which is still the leading cause of death of children in most countries in the world and certainly in Haiti.

The big issue in Haiti during the kind of massive cholera outbreak that you heard about a few years ago was that there is no central sanitation system. And in addition to that, when people were creating kind of temporary hospitals to treat cholera you're having to then move the waste, which is highly contaminated, and what initially caused the outbreak, to another area for treatment. And a lot of times they were finding that this movement to the proper treatment area was not happening, and so you were having further contamination of other areas where it was just being dumped in a totally off of the roadside of place and causing additional problems. So the thought here was, in addition to creating a building that isn't a tent that functions, that when you put that investment in you don't have to replace it every six months-to-a-year, which is typical for temporary emergency efforts, this one would last and could change program if needed, but would have, most importantly, its own on site water treatment facility, and this would treat it down to 99.95% of cholera, so there really is nothing escaping.

And the other reason we had to do this is because this particular location is at the water table, so there was no way to have anything filtered down and be cleaned through that filtration system as you would have on a normal site. And so we worked with a civil engineering group in California to really create a system that collects all the water off of these roofs to supply a huge amount of water demand that you would have for constant cleaning of cholera waste, pails, buckets, chairs, everything. And then all of that waste water would be treated on site before it is released.

So, very quickly just to kind of give some examples, and obviously—the very last one that I had, you can tell me whether I should go on or quit—

Dr. TERBUSH: Let's save this for right at the end. And let's see. Art, would you like to go next? Navy Captain, Art Glynn, whose specialty at this point are energy-resilient hospitals. Go ahead.

ART GLYNN: Good afternoon. Again, I'm Art Glynn. And first of all, let me say, thank you. One of the things that I find that oftentimes my colleagues say, "Art, you're way out there. You're thinking out of the box," I find a very friendly audience here, because the issues that I raise amongst my colleagues, every one of, every person I have engaged with here has a similar thought process. You folks get it.

The bottom line that we have is we are all in this together. Today I'm representing the Department of Defense. Sometimes DoD and NORTHCOM specifically is an enigma. We're the cavalry supposedly. Folks, we are all in this together. DoD, as much capability as we have, will not be able to solve all the problems.

Now, what I'm, as Jim mentioned, I also serve as what we call a NEPLO, a Navy Emergency Preparedness Liaison Officer. Primary responsibility is working closely with FEMA and, frankly, to be the last in and first out. There is no question that DoD is very, very expensive. We do have a lot of capability, but we're also, it takes a long time for need. So if we get advanced warning we can be there and we'll come in force, but it will take some time.

So working in concert with FEMA, Craig Fugate is a FEMA administrator who has endured a great deal, and he has done exceptional job, certainly in my opinion. What I like about Craig is his focus from the very, from the get-go was the first 72 hours, communities, individuals, families, it's up to you. It's up to you. Individual, family, community resiliency. If you can't achieve it, you can't be resilient beyond that that's when you ask for help.

Secondly, this whole idea about victimization, he basically said, "Stop. We have survivors. We have victims or casualties." It's a nuance in our world today. And fortunately everybody I see here, survivors, if the day does happen will be survivors. Also talk about as far as energy resilience working at NORTHCOM and in other capacity with a program or project called SPIDERS, Smart Power Infrastructure Demonstration for Energy Reliability and Security. We say SPIDERS an awful lot.

Let me just boil it down to its bare essence. It's a cyber-secure smart grid. It's a micro-grid. We are now entering phase three, which is to what we call islanding of Camp Smith. So, to be able to provide resilient energy for Camp Smith, which houses PAYCOM and MARFORPAC, obviously is a crucial element for any adversaries we might have in the Pacific, whether it's North Korea or China.

And let me also share with you that we have talked a lot about EMP and certainly that would be a very, very bad day. We have also talked about cyber. I will offer to you that we have been at war in the cyber domain for a number of years. Now, knowing a lot of people will bristle. Certainly wearing the uniform, "Hey Art, you can't say we're at war." The reality, if it walks like a duck, talks like a duck, looks like a duck, it probably is.

Fortunately, what is starting to happen, a lot of what we see behind the proverbial green door, classified, that I'd love to share, but obviously I don't want to go to jail, but a lot of the information that has been coming to the forefront as far as the various cyber-attacks we see are we see power outages. And power outages, to make comment for the folks who were participating yesterday, question that I asked is the various cyber-threats we see, the vast majority, nearly 60% of them have been targeted against the energy sector, primarily the electric grid.

We have talked about what is the day after, one second after, if you will. It doesn't really matter how it happens. The response is going to be very, very similar. The only difference if we had an EMP, obviously wearing the uniform, Number One responsibility is homeland defense. So, oftentimes NORTHCOM, let me segue it a little bit—NORTHCOM has three missions. Number one mission is homeland defense. Secondary mission is theatre security cooperation. And the third mission is defense supports of authorities.

If we are attacked number one mission is homeland defense. So, there is a perspective certainly from a planning standpoint that DoD is going to bring all this capability to the fight, toward the response. That may not be a reality. But the one thing that DoD does very, very well is plan. I mean we do it in our sleep, sometimes too much, particularly when it comes to crisis action planning.

I know in my own family my wife says, "Art, you think too much." Well, it's better than not thinking at all. But I think that's what the group that we have here today, and it was yesterday, is you are all a part of the solution. You're thinkers. And it is a whole community-type of solution.

One of the things that I've thought about yesterday and what we have seen today, one takeaway that I want you to have, and I've said it several times, is we are all in this together, and we have a choice. We can be victims or we can be survivors for our families, our communities, and most importantly for our nation. Thank you.

[Applause]

Dr. TERBUSH: Were we able to get some slides for Terry, for Dr. Donat? Got a thumbs up, IT. Alright. Resilient IT. Terry, you're up.

Dr. TERRY DONAT: Thank you for asking me to come here and speak. And it's a pleasure to serve on this panel with everyone here. I feel guilty because I actually am using the lightest bandwidth, even though everyone else did a really good job in working without that. Looking at communication resilience, I look at the problem as if you look at, if you're looking for what is a resilience model I look at the human brain as probably the most resilient tool or biological formation I can think of that works. It has worked for us for 200,000 years. It provides us capabilities. It has certainly some problems, but while we are still trying to map it out, and that's what these are, these are maps of the main circuits that we're looking at in the brain as to how it works, it's at the gross level, not the micro level yet, we certainly have some capabilities and characteristics of how the human brain works and spans communication between two people that we can look at and try and repeat in our disaster modeling when we don't have the full bandwidth and the full amount of energy we would like to use.

Quickly, I am both a surgeon and a radio officer for the state and in both voice and digital communications, and the reason I did that is I think healthcare is going to fail not because medical personnel are going to leave the scene and flee away; it can fail certainly because you don't have supply lines that are satisfied. We certainly have a saline shortage in the United States right now. But I think it's probably going to fail worse because of coordination and collaboration through lack of communications.

I basically can say the best example I spoke to Mary Laski about over lunch was if you have triage across a state or a region and in one place triage means you get dialysis and in another place it doesn't, it's not the fact that you're triaging, it's the fact that you're doing it unfairly and it's going to lead to a lot disruption and social problems, and also allocation of resources will be a problem.

So, when we look at what are the resilience characteristics that the brain has that we're trying to repeat in communication, one of the things we have to look at is just like your brain, we have in communications a person with a transceiver, meaning you can both send and receive signals has a producer to consumer ratio of one-to-one. Now, in search and rescue, the people who are looking for the person who may have a radio are many people trying to talk to one or the person who is trying to be rescued is one-to-many, but at least you have a one-to-one. And if everyone had the ability to produce energy, produce food, produce water, produce healthcare we wouldn't even have this conference, the need for it. But in a resource limited environment you have a problem where you have to have resources, and certainly for radio, especially the amateur community, you have pre-distributed information, pre-distributed facilities and training. You have it's familiar. You have it's language agnostic, meaning it doesn't really matter which language you use, especially if it's digital.

And it's portable use for responders, their families, and the public. And if you think that you're going to be able to go and work for weeks or months without talking to your family, as well as doing your job, good luck with that.

We have redundant integration over paths. Certainly the digital pathways that are currently used are in maritime, certainly air and terrestrial in space, and through time. Asynchronous communication, store and forward messaging that is allowed through digital communications.

Self-regulating and self-healing. Certainly self-regulating in the sense that people want to communicate and therefore the effort will need to continue that communication, even when you have barriers that you might meet. The other thing, the self-healing, if there is interference or if there is someone who, because of the nature that they have to sleep, can't continue on for 24 hours you have someone else who can step in with that equipment.

It is adaptive through inherent diversity advantageous change, meaning inherently we are all here with different databases in our head. We look at the communication rate. Without having any other background or slides, we were communicating, we speak at about 150 to 200 words a minute. We read at about 300 to 400 words a minute. And we think at about 600 words a minute. Do we need bandwidth in the megabytes per second in order to be able to communicate effectively? I would say if you really believe that then I'm not quite sure what you were doing here and how much information you were actually getting all day, because you were communicating at a far lower rate and getting very effective information from all those people that have been presenting today.

The prioritized capabilities based upon energy availability. We live in a society where we can go to McDonald's and have lots of food or you can live someplace where you have very meager energy requirements, and our activities and our communication through the brain get prioritized, and the same thing is occurring by radio where we can use 12 volt batteries or other sources that are very low voltage to be able to communicate over large areas.

Potential for emergent properties, where the sum of the parts is greater than what you have individually. You may not be able to access databases across large areas, but if in each message at the end of the message you include the weather conditions at your local station someone can propagate that or accumulate that and have some forecasting ability over time. And then the fact that it's active infrastructure Winlink 2000, which you can be, the setups to that will automatically choose to use the Internet if it's available or choose not to and go by radio directly in the case that it's not.

One other thing I just wanted to mention, the structure on the left here, that is the connected diagram of all the major pathways in the brain, red being the most connected and blue and violet being the least connected. While you would think that the brain is highly directly connected, it's actually not. It's mid-range. And that allows for the most adaptable connections between the hundred million neurons and the hundred trillion connections that are estimated.

Just some quick advantages of digital radio. Situational awareness, both internal and external. Hospitals are not small. Anyone who has walked in a hospital, you realize you can't communicate through those walls across many floors if you don't have communications. And hospitals are, in general, radio poor right now. They rely heavily upon the infrastructure. And that's one thing that probably will have to change to some degree.

We are accused in medicine of using fancy and unfamiliar words, and when you start having lists of patients and pharmaceutical terms, medical terms, logistical supplies, even photographs of victims, spreadsheets, checklists, financial records, and personal assignments over many times you're going to have to be able to transmit these very effectively.

100% accuracy for messaging. When people say, "How can that be?" In voice we certainly make mistakes. I'm making some now. Maybe you know. Maybe you don't. But the point being that with digital transmission you have something called a check sum, where if the sender and the receiving station don't match up in the sum of the characters that are there it automatically checks itself and resends it.

You're able to provide both local and long-haul communications and you're able to encrypt messages because most of the systems through Winlink is actually an email server, so you can actually encrypt both your messages and your attachments, which is important. Say for instance you want to have transportation of strategic national stockpile materials; you don't necessarily want to broadcast where those things are going.

Reliable operation on low-power levels and excellent portability. Hastily formed networks came out of the Naval Postgraduate School during Katrina. These networks were satellite-based where they got the Internet, but it beamed it down to 802.16 wireless wide range areas in the city of New Orleans, and then locally at 802.11, and this provided broadband coverage and this has been aggregated in Katrina, Rita, and several other locations through the Naval Postgraduate School.

And just sort of short on time, I think the Naval Postgraduate School site is a great place to go and learn more about these. And that's just the setup. Winlink 2000 I mentioned. This is essentially the radio computer which has a sound card and a modem. You're able to either go directly radio to radio, radio to what is called net radio messaging servers, which are located around the country, and then through the Internet there is common message servers around the world. And this is what Mars and RACES and also the private sector can use in Winlink to be able to act, send messages throughout the world on low power.

I won't go through all the advantages of the digital communications. Much of them I mentioned before, but I just refer you to Winlink 2000. I think the biggest thing is you can send large amounts of data, maybe a thousand or $2,000 dollars sent, tens of thousands of words per minute. So, anything that would possibly exceed the data that you could possibly take care of or interpret in the time that you need. And then the accuracy is important.

And lastly there are two, so besides the private sector and besides amateur radio, you certainly have the two big programs, which are Mars and RACES; RACES is under the emergency management of either your town or your state. Mars is run by—it's a joint private/public partnership between the military, all three branches. And I think one of the parts of resilience is to make sure that when we have programs in place to support those and try and continue them, Navy Marine Mars right now is sort of chiefless, and I think anything we can do to, if there is anyone here to help support and get that program and continue running, because it does involve assets in the hundreds of thousands of dollars around the country that are provided primarily by the private sector. Thank you.

[Applause]

Dr. TERBUSH: Well, thank you, Terry. I've got to tell you, the panel exceeded all my expectations and my expectations were very high indeed. Not only that, but they overcame in difficult circumstances, so thank you again to the panel.

[Applause]

Dr. TERBUSH: Chuck tells me we will be able to see the slides.

Question from unidentified audience member: So my question is for Sierra. I work as a bacteriologist and also a microbe forensic scientist. So, I know you discussed about how you create the hospital, I think the physical structure to be more easy going without electricity, but I am more interested about how in that kind of hospital they were having equipment to identify infectious disease without electricity. I don't know if you could give me some ideas what they were using, because I work for the Greenwich Health Department Laboratory, and we are trying to find equipment, because we're thinking about if the electricity is not working, but at this point we haven't found much equipment that we could use to serve the citizens that are in our population. Thank you.

SIERRA BAINBRIDGE: The short answer is that I, working on the architectural side, don't know as much about the identification of what you're talking about. We have backup generators and that is what we relied on for the most part for some of the higher tech things, especially some of the TB equipment. And they tend to use for most other things very, very low tech, unless it's like these XTRTB, MDRTB identification systems. I can't remember the name of those, that exact machine. But we also haven't found any of those that haven't had that. So in places that we have worked with pregnancy-induced hypertension (PIH), like Lesotho where they actually have to kind of fly these samples from places with no electricity into cities and labs that are higher, have a higher accessibility to the grid, and those are the only things that I have seen so far.

Dr. TERBUSH: The technology that Sierra mentions is imminently adaptable here in the United States, and in fact in many cases we do use it. But it's possible to go ahead and culture bacteria. There are rapid diagnostic tests and other low-energy consumption types of diagnostic tests to what you suggest. I just want to finish with though a quote from Dr. Paul Farmer who is sort of the patron saint of International Humanitarian Assistance, and I just found out that Sierra and her group are working directly for Dr. Farmer. Dr. Farmer, a long history in Haiti, one of his famous quotes was that, "Why is it that the poor always get the worst?"

And I think we can see from some of the technology that Sierra has talked about, some of the innovations from Terry and Don, and then of course our motivator here on the end from NORTHCOM, Art Glynn, I think you can see that even in very difficult circumstances we continue to provide good care at a very high standard. Thank you.

[Applause]

CHUCK MANTO: Thank you to our wonderful panel, showing how we can exploit new technology, even technology we're using in Africa and Haiti, to make more resilient in the United States. Thank you very much.

Growing Inter-Dependency of Gas and Electric Grids

Commissioner Philip D. Moeller, Federal Energy Regulatory Commission

https://youtu.be/k4A4jU5L0WI

Presented at the Dupont Summit on Friday, December 5, 2014

This presentation provides an overview of the growing interdependence of both gas and electric supplies providing a backdrop of information on growing vulnerabilities.

PHILIP MOELLER: Thank you for the introduction and for the invitation to be here today to speak on one of my favorite subjects, which is the challenge of coordinating or better coordinating the gas system and the electric system. I took a look at the registration list. And most of you are new to me. You're not really in the FERC world. So I'm going to take this down with a bit of a history perspective on what we're doing, and then future challenges.

I don't typically title my remarks. But, if I did, I'd probably say the gas–electric challenge—it's a good challenge to have, but it's still a big challenge going forward. I always have to preface my remarks, because I come from an independent commission, that I only speak for myself, which means that you are free to disagree with me if you wish. But I don't speak for the Administration or the agency or the staff or the Chair.

It's important to remember, as part of an independent commission, the Chair of the agency really sets the agenda. In our case, because we're a quasi-judicial organization, when we vote on things, which is typically about five or six times a day on average, it can range from very small re-hearing orders, to multi-billion-dollar pipelines, the Chair still needs to get three votes for it. But, nevertheless, the agenda, in terms of where the agency is going, is set by the Chair. We have a current Chair who's been in office about a year. And we'll have a new Chair that takes over the gavel in April, a sitting commissioner right now.

So, the general big topic of gas–electric coordination. What's been happening, lately you've heard a lot more about it, but it's actually been a trend for, if you look at the data, for about 25 years, that we are just plain using more and more natural gas to generate electricity. For those of you who follow it closely, you know that the most efficient use of natural gas is not to boil water and make electricity. Rather, it's a direct usage on industrial, commercial or residential applications, where you're heating air or particularly heating water. Much more efficient than generating electricity by boiling water, burning gas.

Nevertheless, to me there are five trends or five factors that have accelerated this trend, particularly in the last few years. The first is that, generally speaking, it's easier to build a gas plant than any other type of plant outside of a renewable base. You're well aware that coal is basically, its prospects for new generation are dim at best under the present circumstances. Nuclear, of course, has issues related to the high-capital cost initially, even though the fuel costs are low. And so, except for up rates of existing plants, and those few plants that are under construction in the southeast and vertically integrated markets, we're really not seeing the nuclear renaissance that we thought we might see, say, in 2007, when natural gas prices were so high. So, number one, it's easier to build a gas plant than it is basically any other type of plant.

Secondly, sometimes it's actually electric transmission, which is a more efficient solution to a particular need for power, whether it's trying to resolve congestion issues, or enhance reliability. But it is so difficult, frankly, to build electric transmission in this country, that sometimes those involved, which is not FERC, we don't decide, we don't mandate the construction of power plants. That's not our role under federal law. But, for either the private developers or perhaps the state regulators who are ordering a utility to build something, they will often go to the default of a new gas plant because it's so much quicker than a potentially more efficient but longer-term solution of additional electric transmission lines.

There's a lot of reasons for that. I can go into them. But the fact is, that it sometimes takes the place of a new natural gas plant, or a gas plant will take the place of what logically should have been an expansion of the transmission grid.

The third reason that we're using a lot more gas, to make electricity is that we needed it to back up intermittent renewables, which are becoming a bigger part of the grid, particularly in certain areas, California, certainly it's widespread in Hawaii, now, for those of you who have watched it. The intermittent nature of wind and solar, of course the fuel is free, but you can't count on the wind always blowing. You can't count on the sun always shining.

And particularly in the morning, and more so in California, the evening ramp periods, where the sun is setting, or the sun is rising, but particularly in the evening when the sun is setting, there will be an enormous need for fast-acting gas plants to make up the difference of lost generation when the sun is going down, and, frankly, at the very time people are coming home and demand is up. Only really fast-acting gas plants are in a position to respond to that. So that is a growing trend, which will continue to accelerate in the near future.

The fourth reason that we're trending this direction is that we have, of course, a domestic supply that was almost unimaginable 10 years ago, thanks to the advances in hydrofracking and horizontal drilling. I sat on a National Petroleum Council study for two years. It actually concluded about three years ago. And it was a two-year effort focused on oil and gas production in North America. And, of course, the volume was about that thick if you really want to read it. But the takeaways that I have to distill that two years of work down are, number one, North America has an enormous, enormous amount of gas. And number two, the technology will only allow us to find and extract more if we, as a society, allow that to happen.

The point is, that we can restrict either places that we're going to extract resources. Or we could potentially— I'm not advocating for this—as a society, we could restrict the type of extraction methods. But the bottom line is that we will only find more, we'll only be able to extract it more efficiently. So, if you can't tell, I'm obviously bullish on the long-term future of gas in this country. You know, you might like it. You might not. I'm trying to make an observation in terms of how it fits into planning and what we're talking about on this general subject.

The fifth reason that we were accelerating more gas, as was alluded to in the introduction, the fact that we have a suite of environmental rules that range from cooling water intake to how you dispose of coal ash to, of course, the very obvious air regulations that affect not only mercury and air toxics, the rule that's presently being implemented, but of course the clean power plant that's been suggested by the EPA, all of which, to some extent, are to the disadvantage of more coal production. They're not all exclusively aimed at coal, but for the most part, they will accelerate the trend toward basically burning more gas to boil water and make electricity from that.

Again, I'm not putting a value judgment on it. Rather it's an observation that, whether you like it or not, these trend lines have been pretty clear. And to me, they've been particularly clear over the last five years. But the event that really caught my attention, and it developed—helped generate the concern that I've been expressing over the last few years, was the February 2011 outage in the southwest. You might remember, it was right around the Super Bowl that was being held in Dallas. There was unseasonably cold weather, but certainly not unprecedented cold weather, affecting Texas, New Mexico, and Arizona.

And, what resulted from that was a variety of incidents from well freeze-offs to poor communication between the pipelines and the electric utilities. And I believe it was about seven million people in Texas lost power, a good 20,000 meters or 20,000 households in New Mexico lost their pilot lights when it was very cold. There were outages in Arizona as well. And it was a big event that had a lot of consequences.

And subsequently, our agency, FERC, along with a representative from NERC, the North American Electric Reliability Corporation, put out an outstanding report on the causes of it, and 34 recommendations, most of which were focused toward state legislatures and state utility commissions. I think there were six on the electric side, maybe eight, the rest on the gas side. And it was one of those great reports. But it also happened to be released in August of 2011. And you know, for those of you who live in Washington, you know that if you release something in August in Washington, you better have a pretty good rollout plan for people to notice it, because a lot of people leave town. And their focus is somewhere else.

My concern was that this was indicating a series of actions that could be replicated throughout the country as we grew more dependent on natural gas from our traditional, let's say, fuels of coal and nuclear. And I'm from the northwest, so I'd put hydro in that. But that's not as ubiquitous around the country as it is in the Pacific Northwest. And I was hoping that this was going to generate a national discussion on this growing dependence, because if you read the report, and I certainly recommend you do, because it's a good primer on the gas industry, on the electric industry, and the discussion of what happened, and the recommendations, you realize that these are two very different industries that are converging. They're regulated differently. They're financed differently. Electricity moves at the speed of light. Gas moves through pipelines at 20–25 miles an hour. The production of the base fuel is different. For instance, we have citing authority over natural gas pipelines across state lines. But we don't have an equivalent authority on the electric side. So those are cited state-by-state or county-by-county. Two very, very different industries.

And most of the time, people have grown up either on the electric side, or on the gas side. Most utilities, with a few exceptions, are either electric utilities or gas utilities. And sometimes the same word means something very different to each industry. And so the concern was that these industries are converging. They don't know each other very well. And there are profound implications. When you go from a system where you can rely on a 60- or a 90-day pile of coal, recognizing that sometimes, in winter, coal piles do freeze, when you have a set of fuel rods that only get changed out every 18 or 24 months, that's a very different dynamic than being more reliant on a just-in-time fuel delivery which is what we are increasingly becoming dependent on as we burn gas to make electricity.

That's where the resiliency issues kind of come in. My frustration, then, manifested itself in a letter that I just kind of put out to the world, roughly in February of 2012, almost three years ago, saying, "We've got these big issues." My concern, based on the outage, I think needs a more thorough discussion of, is there adequate communication right now? Are there inefficiencies in the fact that the market times of each industry are different throughout the country? How do we deal with the long-term issues of financing new pipeline to meet these needs? What should we do about all this?

And, to his credit, our Chairman at the time, once he saw the enormous response that this letter generated, we actually got an official FERC docket number assigned to the letter. And then, one of my colleagues added a few more questions. And thus we began what's turned out to be a multi-year effort to further examine this issue and hopefully come to some solutions, although we've made a lot of progress, we still have a long, long ways to go.

The first thing we did, we had five technical conferences focused on the various markets. Because, if you don't follow this closely, you may not realize that we have parts of the country that are in very sophisticated electric markets that involve a day-ahead market, where you bid in, and you may or may not be called, based on your economics of your plant, and then a real-time market, that deals with the real-time variations that differ from what was projected the day before.

We also have significant parts of the country, notably the southeast, and much of the inland west, which is a typical bilateral market, oftentimes vertically integrated, where the same entity owns, on the electric side, the power plant, the lines to transmit the power, and the lines to distribute the power. That's different than what we call the more organized markets, where essentially, the generators earn a class by themselves and compete against each other. And then the load-serving entities actually deliver the power through the transmission distribution lines.

That's the way it is. But it does kind of complicate how we would deal with, say, national solutions to this. So we had five regional meetings focused on the different regions of the country. These were held in the summer of 2012. And at the risk of potentially offending some people, I would say that, at that time, generally speaking, the pipeline industry didn't really feel that this was a very significant issue. And I'm happy to say that I think that that has changed in the last two-and-a-half years.

My nervousness kind of continued because, if you don't count last winter, you may recall that we had two extraordinarily warm winters in a row. And my gut feeling is that we, as a nation, and as regulators, and as people involved in the industry, we're kind of being lulled into a sense of complacency, because the system wasn't stressed, frankly, the two winters prior to this one. It was pretty easy to get through. And, in addition, a lot of the power plants that are going to be closed down under the mercury-near-toxics rule hadn't yet been closed down.

Fast forward to last winter, as we were approaching the winter, we finally came up with a proposal in a rule that said that the pipelines and the people who run these organized markets, which covers most Americans, we made it very clear through the rule that people could talk to each other. It would be voluntary. But if there was some kind of a weather event coming, or there was some concern over a sudden loss of a pipeline that was, perhaps, fueling several power plants, that that communication was allowed, encouraged and legal. There had been some concern that, because of sometimes the sensitive commercial nature of it, that people could be violating the law by telling a market operator that there were pipeline constraints that weren't publicly known.

Fortunately, and I will give OMB (Office of Management and Budget) credit to this, they expedited the effectiveness of our rule. So, instead of a standard 60-day, we cut it down to 30. And it became effective, I think, December 23rd of last year, which turned out to be pretty good timing, since the polar vortex event Number One hit on January 5th, and the system was extremely stressed. There was never a proposed—Nobody lost their power. But it was razor-thin close to there being some load-shedding when it was extremely cold.

Now there were a multitude of factors for that tightness. Yes, some coal piles did freeze. And a lot of natural gas plants without proper weatherization didn't work. Some of those plants that had the dual fuel capability of fuel oil found out that, because of the temperatures and the lack of, perhaps, some heating elements, the fuel turned to gel. And they essentially were inoperable. So there was a lot of the fleet that was down for reasons related to those issues I've gone through.

There were also a lot of plants that didn't have the gas that they thought they would, because the demand was so high on the direct usage side from local distribution companies. That was a major wakeup call. There will be ramifications from last winter including filings before us now, that I can't talk about, trying to enhance the reliability of those units that have been part of the capacity markets and are paid to provide capacity. And, that's a controversial set of issues that will be debated on paper over the next few months.

But my takeaway was that PJMwhich operates the biggest market, arguably, in the world, went into last winter thinking everything was fine. *(PJM is a regional transmission organization (RTO) that coordinates the movement of wholesale electricity in all or parts of 13 states and the District of Columbia.)* They actually told us that. And then, when they skated through three very tough polar vortex events, found out the system was not as robust as they thought it was. And it has shaken a lot of people's faith in the reliability of the system, or at least under the status quo.

In the meantime, we've been doing another major effort to try and create more efficiencies in the way these markets operate. The gas market starts at nine a.m. Central clock time, which is different than when the electric markets start, which are typically a few hours later. There is a natural inefficiency there, particularly based on the present nomination cycles, where sometimes, if you own a power plant, you have to—you're bidding into the next day market, not knowing whether you're going to be called or not. And there's a big question mark as to, do you go out and secure a firm supply of gas, not knowing if you're going to use it or not, based solely on when these clock times occur for markets to begin? That is a natural inefficiency.

And we proposed solving it by moving the gas day to four a.m. Central clock time. And we achieved something remarkable. We achieved unanimity in the gas industry. They don't like it—at all, universally. FERC has brought the entire gas industry together in opposition to our proposal. That is a pending rule. So I can talk about it, since it's a rule. But there are other elements of the rule, particularly we asked the North American Energy Standards Board to take a look at it. They obviously couldn't reach consensus on when the gas day should begin. But they did reach consensus on adding a couple of more nomination cycles to the gas day, which should certainly help if they were to be adopted, something that we'll be considering. Of course, that won't do any good for this winter. That would be toward the future.

Kind of the third set of long-term issues, which are particularly challenging, and all creative ideas will be welcome, is how do we finance new pipeline infrastructure? Pipelines have traditionally been financed through long-term contracts with local gas distribution companies. They, of course, must have gas whenever they need it, because the last thing that they want, at all, is to lose gas and have to go around manually and relight thousands of pilot lights. So they put a premium on having gas, and the absolute highest demand time imaginable, based on historic and projected usage.

So they have been willing to pay, basically, the higher cost of firm supply. That has essentially allowed, in many cases, along occasionally with industrial consumers, to provide the financing background for new pipes to be expanded. That's traditionally how it's happened for decades and decades in the United States.

Gas consumption, at the residential level, is basically flat, even though customer bases have increased. We have much more efficient appliances now that have, despite the growth, led to a relatively, depending on the area, relatively flat demand going forward in terms of gas consumption there. The new customer base, the people who are really using the big amounts of gas supplies now, and more so going forward, are power plants. And yet, most of the power plants are in these markets where they are competitive markets. And consumers benefit from the fact that they have to bid in, and they may or may not be called based on their bid and their costs.

But they're different than the traditional base load plant that's going to be on 24/7, roughly 365, except for maintenance. And so they are in a position where they would argue that it's financially impossible for them to sign 20- and 30-year contracts, because they don't know, on a day-to-day basis, whether they're going to run. If they don't run, they don't get paid. And yet, this is the group of new customers that are requiring a pipeline expansion.

So we've got a real conundrum as to how we can get new pipes financed. It's a real issue right now in New England, given the fact that they are becoming increasingly, in a relatively short amount of time, 10–12 years, reliant on gas for electricity. And essentially, they have been shutting down their fossil plants, the nuclear plants, at least Vermont Yankee being shut down, not huge but important from a load pocket and from a cost basis. Nevertheless, is getting shut down. It adds to the concentration of gas as part of their portfolio mix.

The pipes through New England, you know, kind of at the end of the line, are basically fully subscribed. And there's almost universal recognition that new pipeline is needed in New England more so than anywhere. And some of my environmental friends do not agree. They want the existing pipeline infrastructure used even more efficiently, and are concerned about a 30- or 40-year investment that then will drive policy toward more gas. Nevertheless, almost everybody else admits that this is a real problem.

My concern is that, if we can't solve it for New England, and I do give credit to the governors for coming up with a solution, although it has some challenges related to federal law, at least they came up with a creative solution. But, if we can't get it solved in New England, my concern is that this can again be replicated, both in PJM, perhaps in New York, and certainly in my cell, as the clean power plant has a very aggressive goal, at least in one of the building blocks, of us going to 70 percent dispatch of natural gas plants. That would require a lot more pipe. And I don't know how we get it financed under the present system.

So creative ideas are welcome. And that's one that's going to be with us for a while. And hopefully, there won't be problems before there are solutions. But, we're in a very tight situation, particularly over the next couple of years, because of the number of coal plants that will be retired under the mercury rule. And you know, if we have mild weather, we can probably get through it okay. But there are different challenges in the winter than there are in the summer.

But we're in for kind of a wild ride, I think, for about the next 24 months. And I appreciate all of you being attuned to it. We are trying to address it. But hopefully, the context or the substance of my remarks have indicated that there are a lot of different interests here, regulated differently. And the solutions are not simple. They may not be elegant. But we're at least giving a good effort. But these are important matters. And I appreciate everybody being attentive to them, and hopefully adding to the discussion in a proactive and positive way.

Thank you for your attention. And, as appropriate, I will stay for questions and answers.

CHUCK MANTO: First of all, a round of applause.

[applause]

CHUCK MANTO: We definitely want to have some time for Q-and-A, because the audience here has been predominantly looking at high-impact threats of any kind to infrastructure, such as the power grid. So you have more prayer going up on your behalf from this room than probably church, except for maybe the possibility the atheists, who get praying when they're really worried about something.

I know the concerns here are not only what's happening to all of the issues you've raised about gas and electric interconnections, the loss of coal plants, and then you throw a monkey wrench into it like some other kind of disruption to the system by a high-impact event. So I know there's a lot of concerns. I'd like to make certain that we have questions from the audience who can identify themselves and raise a succinct question. And I see Mr. Tom Popik has a hand up. Identify yourself anyway.

Q: Hi, Tom Popik, Foundation for Resilient Societies. On behalf of my group, we'd really like to thank Commissioner Moeller for his extraordinary leadership on this issue. Partially due to the interest that FERC and, in this case, NERC and their good work on this issue. Our group has been engaged on what is going to be a very important issue going forward.

So you spoke to the situation in New England and other places of the country, where the home heating gas is under firm contract, whereas the electric generation is in the spot market, as it might be called.

PHILIP MOELLER: Right.

Q: And generators bid into the market, if they can secure the gas supply. If they can't secure the gas supply because there's a cold snap, then there may be an outage. But it seems that the cost of that outage is really externalized to the society. There is no commercial entity, really, that is held responsible under the current regulatory system. I wonder if you could comment on that, and if there is any alternatives that might produce a greater reliability of electricity in these kind of circumstances.

PHILIP MOELLER: Well, thank you for the question. And thank you for your involvement in this issue. It was in your materials when we visited a while back. And I appreciate the attention you gave to it. Because it does need as widespread attention as possible, so that we can come to some better solutions than we have now.

The people who run the markets have basically been trying to address the issue, Tom, by—and I'll speak very generally, because we'll have some pending matters before us—But essentially making sure that, if you perform when you're called on, you perhaps will get, particularly for over-performance, you'll get rewarded. If you do not perform when you're called on, you will be penalized quite severely. And the penalties then flow to those who did perform. Essentially not universally, but in a lot of these markets, the capacities being paid to perform, it's a capacity market where essentially they're buying an option on, "We're going to pay you to be there. But then you got to be there when we call you."

And that has generated a lot of discussion. Some of those changes, though, take several years to implement, based on the capacity market design. And so that has played out in various markets and is in various states of play. That's part of the solution. Hopefully, creative approaches toward essentially firming up gas supply can come out of all these discussions. I don't know exactly how, exactly how it could be done. But I think there's some potential there.

Some of the biggest challenges we have are those gas plants that are behind the city gate of a local distribution company, because again, the LDC is going to make absolutely sure they have enough gas for their end-use customers. So that appears to have been one set of plants that were particularly vulnerable under the present system.

In many cases, the actual generators, in response to what happened last winter, have incorporated dual fuel technology into their plants at great expense. But worth it to them in the grand scheme of how they're compensated, either through a capacity market or through an energy market. And that's a very positive trend.

145

Of course, whenever you burn oil, you're very limited, depending on where you are, to typically how long you can burn it, and how willing the state environmental agency is to give you a flexible permit. But most of those permits are limited in the number of hours they can be run each year.

But the interesting thing about New England last winter was that oil was in the money a significant part of the time, because gas prices were so high. And to maybe add a little exclamation point to my discussion of New England and the lack of capacity, or as some would argue, the need for additional capacity, the highest gas prices in the world are the futures for the winter months in New England, higher than Japan, higher than Asia. That should tell you something. And it's 100 miles away from some of the cheapest gas in the world, in the Marcella Shale.

So I hope that highlights the challenge of the opportunity, is that we have all this domestic gas, but we've got to find a way to finance the pipeline to get it to the market that is begging for it. That is a tough problem to solve. So all creative ideas welcome. But thank you again for your attention to this subject. And I'll point everybody as to the report that's available on our website, where if you're kind of afraid to ask for a primer on the electric side or the gas side, it's a nice little primer that was issued in August of 2011.

CHUCK MANTO: Identify yourself please.

Q: Commissioner, I'm Frank Gaffney with the Center for Security Policy and the Secure the Grid Coalition. You are obviously addressing a specific issue in your remarks here today.

PHILIP MOELLER: Yes.

Q: I think many of us in the room are seized with the issue, not so much of reliability as resiliency. And I just wonder if we could take advantage of your being here to hear what you believe the FERC is doing, needs to do, will do if you can speculate on that score. Because, just to pick up on a point, one thing that occurs to, I think, some of us is, wouldn't it be a good idea, if you're concerned about resiliency in an environment in which we may really have this upon us as a crisis at any time, not to close these coal-fired plants? And is that something that FERC is going to engage in as an intramural/interagency exercise if it is as important as it seems it is, from what you've just described, in terms of shortages and real possible survival problems for people in a very cold winter?

PHILIP MOELLER: Well, this is not going to generate any headlines for people who have been following things that I've been saying and doing on this issue. But my concern all along, and it's reflected in multiple times in front of Congress, was not what the Environmental Protection Agency wanted to do with the mercury rule, but rather, the timeline in which they implemented it. I would have added a couple more years. And I think that would have given us added insurance to get through it.

Since a lot of these plants were on the margin anyway, some very old—a lot of them very small. But, what I try to emphasize, and I'm sure it's where they provide grid stability in a way that the grid is engineered, they could provide coal plants inertia, which adds to the reliability of the grid, that you wouldn't have if you just replaced it with a gas plant. Severe concerns about the timing of MATS *(mercury and toxic standards)*. And we'll see. If we have mild weather, we'll probably get through okay.

But, what we have done, is MATS has an additional year. We're starting to see the requests come in for essentially that extra year of plants being needed for reliability purposes. But here is the challenge. Those people who own the plants, they have to decide which federal law they're going to violate. Are they going to violate the Clean

Air Act or the Federal Power Act? And again, I testified to Congress saying, "You really need to fix this so that you don't put the generators in what a former colleague said, the Hobbesian's choice of which law to violate, because you open yourself up to civil litigation if you do that."

So we got a couple in front of us to do that. And we've granted at least one to date. We do have kind of a formal process with EPA on weighing in on the reliability implications, which we did at our last meeting. The more challenging issue, I think going forward, is the clean power plan, because it is so significant. And, in my mind, and I've said this publicly, it's a challenge that they are trying to fit what I would say the square peg of state implementation plans into the round hole of what is interstate commerce. This is an interstate grid. And that is inelegant at best, but potentially very expensive, with major reliability implications.

And I submitted comments to the EPA on Monday, essentially saying that. And it's a proposed rule, so presumably they will consider changes. But very, very concerned, given how front-loaded the requirements are for 2020. But, as I said in the beginning, these are my views only. I'm not afraid to share them. But I don't speak for the agency. And whatever agency actions we take will be up to our Chair at the time.

CHUCK MANTO: I have a quick question on the abundance of natural gas, especially in light of fracking. And one way to bound the problem, from a layman's point of view, would be something like this, but maybe you'd come up with a better way. So let's say all the fracking gas is on the left side of the room. And all the other gas is on the right side of the room. And let's say all the normal sources of gas, aside from fracking, on this side of the room, goes away forever. And all we had left was fracking gas only, for 100 percent of our need for natural gas. How long would the fracking gas supply last if we used it up as much as we possibly could, because there's no other gas, not a whiff of gas anywhere in the world, except for fracking gas? How long would that fracking gas supply last if it were the only gas supply available at all?

PHILIP MOELLER: Well, I think we don't know, yet, because it's a relatively new industry. And so the profiles of well production are still being formulated, although generally speaking, they tend to have a very high-initial output that tails off, but probably a very long tail. And a lot of it's going to depend on new areas. For instance, California has a significant potential for fracking and gas there. But, will it be allowed socially? We don't know that yet. I'd probably make the same argument with New York.

However, in the areas where here are fracking, where there is fracking, the Marcellus numbers have just been unbelievable. And the next wave is the Utica Shale in Ohio, roughly. And the projections there are stunning, 500 percent.

CHUCK MANTO: So would that be 30 years of supply? Or 50 years of supply? What does that mean?

PHILIP MOELLER: Well, there's a lot of fracking gas out there. And there's no reason to believe that the other areas, although they're declining, would go away. Plus, we've got the potential of Alaska, which is just enormous. But, of course, getting it from the North Slope to other places is the challenge. But I'm not worried about the gas supply. I think we've got—like it or not, I mean I'm not trying to weigh in on being pro-gas or anti-gas, I'm just trying to call it as I see it, which is an enormous—.

CHUCK MANTO: Like 100 years or 200 years or 500 years?

PHILIP MOELLER: I can get you the EIA (Energy Information Administration) numbers.

CHUCK MANTO: There you go. Okay. Great. That'll work.

PHILIP MOELLER: Something that we won't have to—our grandchildren won't have to worry about.

CHUCK MANTO: The grandchildren won't have to. Okay, that's fair enough. We have another question. Identify yourself please and ask your question.

Q: Yes, Terry Hill with the Passive House Institute. Along the lines of the gas, how many years, there's a tremendous amount of inherent energy embedded in existing buildings. Is any thought given to trying to capture that in your plans?

PHILIP MOELLER: Well, there is a lot of effort going on in innovative ways to, number one, get building managers to manage energy better. And those are often state programs, and sometimes privately run programs. Innovative thoughts about how you chill water at night, particularly in the summer, and then take advantage of that cooling during what otherwise would be peak times during the summer. That's really not in our realm, other than we have the power, in some cases, to create market structures where that could be bid in. But it's not something that we would mandate.

We have been promoting, particularly under our former Chairman, demand-response. You know, people getting paid to not use electricity. I have always been a supporter of that. But my support has been qualified as to how much you pay. And I was the dissenter on our big order. The order would have allowed demand-response to be paid the full locational marginal price, just like a power producer. I argued that you should subtract the cost of generation from that compensation, because otherwise, you got to pay for the generation to be there. It's not fair for them to not have that backed up.

The courts backed me up, but the courts went even farther and said we didn't even have authority in this area under federal law. So now that's being—the Agency is asking the Supreme Court to look at that. But that's an area that kind of falls into the demand-response, in terms of you being able to bid that savings into a market. But, for the most part, that's going to be handled at the State level.

CHUCK MANTO: As Ambassador Cooper comes up, and Andrea Boland, we have one more question. Please identify yourself.

[00:44:03]
Q: I'm David Bardin. As I mentioned earlier today, my first civilian government job after I got out of the army was in the predecessor agency to Commissioner Moeller's agency. I joined them in 1958 and worked there for 11 years. I've really got three questions, Commissioner, all related to the authority that your commission now has under Section 215 of the Federal Power Act, which didn't exist when I was there.

First has to do with standards and exceeding standards. We have a complicated process that I won't go into, in which the FERC approves standards after the National—NERC has recommended standards, perhaps at the request of the FERC, perhaps not. An earlier panel today, we had a former Assistant Secretary of Defense who told us that a number of electric utilities are going well beyond the mandatory standards to do more than the standards require, because they feel it's appropriate in terms of their service area, their customers, their sense of mission, whatever.

So my first question is, whether you as one of the five Commissioners feel any discomfort at the idea that one of your regulated electric utilities, or ideally many of them, would exceed these minimum standards the commission sets.

PHILIP MOELLER: No, I wouldn't feel uncomfortable about that. I mean if you take—let's take one set of security. Some areas of the country are more important financially than others. So perhaps they should exceed standards.

Q: The second question has to do with these high-impact risks that we talk about, and that you've talked about, which is what we've been discussing today and yesterday in this process, the things like cyber attacks, and malevolent attacks, and maybe a malevolent EMP attack. Maybe none, just natural event, of space weather. And there's a question of just information, before you get to the question of analysis and what you do about it.

And talking with one of the researchers who was here earlier today, but has had to leave, he has a concern that I want to share with you and see whether you have any tentative thoughts on. The background is this. The Electric Power Research Institute has a data collection program, which I think is wonderful. I believe they have 37 nodes around the country. They and I certainly wish there were a lot more. They've got the cooperation of some utilities. They work with the NERC's geomagnetic taskforce.

But the complaint that this researcher was making, and I don't know how valid it is, is that the data that are collected—and this is really empiric data on geomagnetic-induced flows in our systems, these data are not being made public. Somebody feels—calls them classified. So my questions to you are, one, has the FERC required anybody in the industry at either of these groups to withhold the data from the public generally, and the research community specifically? If it hasn't required it, has it authorized it explicitly or implicitly by your periodic reviews of the NERC programs? And do you have any personal opinion as one of the five Commissioners as to whether such empiric data should be generally available for the research community to dig into, analyze as best they can, and argue about?

PHILIP MOELLER: Well, I'm not aware of us restricting access to that data. I don't think we've been proactive on that matter at all. EPRI, as you know, is a member organization. So, to the extent that it's proprietary data that they don't feel comfortable, or maybe by their bylaws cannot release, I don't know the specific details, but that wouldn't surprise me. I'll look into it and see what the status is. Sometimes you want this to be public. Sometimes you don't. And sometimes you want to make sure the right people have it, so that good decisions can be made. But I'll look into it.

Q: That would be great, Commissioner Moeller. I can't ask for more. When Commissioner Moeller says he'll look into it, that means a lot. My final and quick question is this. The natural gas pipeline industry is also a consumer of electricity.

PHILIP MOELLER: Yes.

Q: As are the crude oil pipelines and the product pipelines. And are there aspects of the oil industry, the upstream production and the downstream refineries, they can't run without electricity? Has the FERC had occasion to look into the specific questions of reliability, vulnerability, resilience of power, electric power supply, to these other critical energy industries?

PHILIP MOELLER: Well, the Southwest outage of 2011, some of the more extreme examples were, because of the situation and not enough gas, and the need for rolling blackouts, turned out areas—they had load-shedding blackouts in areas, and didn't realize that, by blacking out this area, they also knocked out the compressor at a gas pipeline, which made, obviously, things worse. And so that region assures me that they've learned the lessons there. And I think we saw something similar with the polar vortex events that affected most of the eastern interconnection last year.

A renewed need to have the electric side understand the gas side, so that they could designate a compressor station is something that doesn't get blacked out on a rolling blackout, and to increase the communication, and essentially identify those critical structures. Has it been a massive effort by FERC? No. But I think we've helped add to the need for the entities involved to actually talk about this and plan for it in the future.

So I'm feeling relatively comfortable that, through some people's pain, lessons were learned. I appreciate the attention today. I also want to introduce Robert Ivanauskas who is an attorney on my team. Some of you have worked with him. He's done a lot of these issues. He's here today. Thank you again for inviting me.

[applause]

CHUCK MANTO: Thank you. And, on behalf of the FBI sponsored InfraGard, we wanted to thank FERC for all the work their staff has given us over the years, helping interact with these folks from across the country. Here is a little token from the DuPont Summit. And this is our new book that's a multi-media presentation of these issues. And we thank you for your participation today.

PHILIP MOELLER: Thanks very much.

Next Steps for States and Local Communities in Light of Vulnerabilities and Limited Federal Remedies

https://youtu.be/Ksus-g2e2_Y

Presented at the Dupont Summit on Friday, December 5, 2014

Mr. Bill Harris, International Lawyer and Secretary, Foundation for Resilient Societies
Mr. Tom Popik, Chairman, Foundation for Resilient Societies
Honorable Ms. Andrea Boland, former Maine State Representative
Ambassador Hank Cooper, Chairman, High Frontier

This panel covers the efforts at the state level given slowness of federal level regulation and guidance to meet higher-impact threats.

TOM POPIK: Okay, thank you. So I'm really pleased to introduce our panel today, starting out with Andrea Boland, who was recently a state representative from Maine, and the author of a really pivotal bill for our movement, an EMP protection bill for the State of Maine, which ended up causing a very extensive study of EMP protection for Maine, and really, I think, has national implications. William Harris, who is the attorney for the Foundation for Resilient Societies and truly an expert in both federal and state regulation of electric grid reliability. And then Ambassador Hank Cooper, who has a really unique experience base in the State of South Carolina, in regards to National Guard involvement in resiliency, including resiliency of the electric grid. So that's our panel today. And I think we're going to have some really interesting things to say.

I believe, first in the lineup is Bill Harris.

BILL HARRIS: Thank you. So we're running late on the schedule. I will just try to give a brief overview of some opportunities and synergies between federal and state regulation, why more can be done at the state level, sometimes at the local level, and in compacts among states. So I had some view graphs. I don't know if they're available.

BILL HARRIS: So basically, one key slide I just provided is a reference. And it will be in the materials I believe of the proceedings, was just the savings clause of the Energy Policy Act of 2005, at which basically—I won't quote the whole thing, but it basically says, "Nothing in this section shall be construed to preempt any authority of any state to take action to ensure the safety, adequacy and reliability of electric service within that state, as long as such actions is not inconsistent with any reliability standard, except that the State of New York may establish rules that result in greater reliability." I'll stay away from that for now. But that does provide a special opportunity for experimentation in New York State.

So the states have an opportunity, when they permit a power plant, to consider safety or reliability or adequacy. And adequacy could be adequacy given some contingencies that are adverse, like the gas–electric interdependence, or the loss of capability in a solar storm, that sort of thing. So that's—Then I just point out, there's an Edison Electric Institute Handbook available online called *State Generation and Transmission Setting Directory*, etcetera. The last edition is October 2013. So you could look online for your state to see what you can do there.

I'd also point out some states give special authority to local municipalities. So Colorado, Maine, New Jersey, Texas, and Vermont give extra authority for municipalities. So a state, a municipality might decide, if there's a new transmission facility going there, that they want a geomagnetic-induced current monitor, if it happens to be in a site that it's possible they could require that.

If there are new renewable resources that are connecting to the state's transmission system, they could require a GIC monitor, for example. The federal government is not requiring them in the current proposed NERC–FERC standard. But they could make a difference. And usually, having the monitors is the first step toward deciding whether you want to buy the protective equipment. If you don't know the hazards at your site, you're most unlikely to buy it. So that's one area where states can act.

Then, following up on what Commissioner Moeller said about the interstate gas pipeline system, the federal government has the authority to extend these interstate gas pipelines. But they happen not to have any authority to regulate the reliability of the interstate gas pipeline system. It turns out no one in the federal government does that. So there has not been a pre-emption by the federal government of state authority.

So a state that is converting rapidly from coal to natural gas, a state like Colorado, for example, might decide that they will not permit expansion of capabilities if there are not requirements for fail-over-gas–electric compressors. Because right now, about 6 percent of the nation's gas compressors are electric only. And people didn't care in the past, and it's a relatively new development. But it turns out, if you lose your electric power, then you lose the gas flow to power your electric plant, and you get a downward spiral.

So promptly, we'd be heading toward about 10 percent electric-only if there weren't regulations. But there are now products where you can buy gas–electric fail over systems with smart supervisory control and data acquisition (SCADA) devices that will actually pick the lower cost for the particular time of day and the particular season of the year. So these fancier compressor systems pay their own way, but they're not being implemented at a fast rate. And if we're concerned about enhancing reliability, that's an example of what a state could do. A state could just require more reliability in this state.

Then, let me turn to the clean power plan, briefly. This is a plan that was proposed in June 2014 by EPA. And it will force a more rapid transition from gas—from coal-fired electric generation to natural gas, on top of all the other transitions happening anyway. If we have that happen, we're going to be losing a large share of electric generation that has the fuel onsite. We basically have nuclear with fuel onsite. And we've had coal with fuel onsite. And we're going to, just in time, long-distance delivery of natural gas with pre-emption by heating customers ahead of the power generation.

So this is an opportunity for the states, but also for regions. So it probably—The studies that have been done in the last six months show that if the states work together through regional planning, and EPA might allow them to do this, they can allocate which power plant transitions away from coal. And they'd need not do it state-by-state, they could do it by region. So they could end up with higher reliability and saving billions of dollars at the same time.

So there's an example where states have authority to act on their own if this plan is implemented. Or they can work in a group. So the regional greenhouse gas initiative has nine states. Pennsylvania may join them. And they've saved a lot of money and improved their energy efficiency by working together and having an emissions trading market.

So I'd just briefly mention, there are some areas where there's federal pre-emption, deciding of liquefied gas facilities, the licensing of nuclear power plants, and the expansion of the gas pipelines. Those have federal pre-emption. But generally, states have much more authority than they have been utilizing.

So the standard under the Federal Power Act involves safety, reliability, and adequacy. But under many state laws, they only refer to safety. And some of the state laws don't refer to reliability. So the states could just change that on their own and give reliability a higher value in the state decision-making process.

So I won't take any more time, because we're running behind. But the basic message is, there's a synergy—a last thought. The synergy between the states and the federal government is shown in Maine, because Maine and Representative Boland and her colleagues demanded that Central Maine Power produce the records of geomagnetic storms monitored at Chester, Maine, and the outages. We have more than 20 years of data which is now helping us improve the federal standard, which is under consideration. So it's not just that the Feds help the states, the states can help the Feds. Thank you.

[applause]

TOM POPIK: So Bill Harris really had some excellent remarks. He obviously has a deep understanding of some of the regulatory and legal issues. I think one of the big takeaways is that electric generation is regulated at the state level. And because you can't have a functioning power grid without generation, there is a really important role for people who are active at the state level to make sure that generation is more reliable.

We also have this issue, as Commissioner Moeller so amply talked about, that more of the generation is becoming gas-fired. And a lot of that gas comes through long pipelines. And some of the states are at the very narrow end of these pipelines. So I'll give you several examples.

The State of California, at the end of a long pipeline, the State of Florida, which is basically a peninsula, also at the end of long pipelines, and then the New England region. Some of these regions are also at the end of long electricity transmission lines, too. And when you look at the combined risk of long electricity transmission, plus electric generation at the end of long pipelines, you can get a compounding of risks, which, in some cases, is quite extraordinary.

So, for example, the State of Massachusetts is 81 percent dependent on long-distance transmission. That's interstate transmission or, alternatively, electricity generation using interstate natural gas flows. So I think that that is a really excellent example of this so-called just-in-time energy delivery.

The takeaway here is, we don't need an EMP event, we don't need a big solar storm, we don't need a cyber attack, and we don't need a physical attack either, for there to be a catastrophic blackout if some of these issues aren't addressed. So, as we go through each winter going forward, all of us in this cause will become increasingly important in terms of both electric reliability and, as Frank Gaffney talked about previously, also, the issue of resilience.

And so now I'm going to turn it over to Andrea Boland, who has been very active at the state level. And I think she can talk about the detailed reality of some of these issues at her state's level.

ANDREA BOLAND: Sorry to make two appearances here before you today. But continuing on from what I had said before about our concerns with the NERC influence on state policymaking, and it being quite disappointing because their standards, their profile is quite low as far as an event, the power of an event would be. So they want to just have to meet a standard of, say, maybe a D-plus if you're in school. And we're hoping for at least a B-plus. And that's kind of been a problem.

One of the things, though, that I wanted to just mention to you, and I'm really feeling kind of disabled by not having the slides here, because I'm not sure how well my memory will serve me. But one of the things that I noticed, and of course I'm not a scientist or an engineer or a physicist or any of that, I've just been trying to follow it as best I can, which is a trick.

So I've relied on folks around me to help me with that. But you heard Curtis Birnbach earlier allude to John Kappenman, who is considered perhaps our best space weather analyst in the country, certainly highly knowledgeable about—Oh great. Okay. So I've done my introduction. Here is what I was looking for.

I just wanted to show you a few slides. So the work that he has pointed out, he and Curtis Birnbach who you heard from before, in some of their work has shown that they've taken—they're pointing out the work of other national experts who have taken information and guessed what—I shouldn't say guessed—they really take pride in being able to validate what they show, whereas we don't see that happening with NERC, in a credible way, anyway.

In the incident of a couple of different storms, showing the threshold to which these storms rise, being 4,000–5,000 nanoteslas per minute, which I'm not going to try to explain. But it's a measure. We can all understand pictures. Whereas, this is what NERC's storm profile would be. This is what we could expect to experience in the red. And this is what NERC would like us to take as a satisfactory goal here. They don't use real data, it doesn't seem, at least that we have been able to have explained to us in Maine. And then, the next slide—.

So what these experts do is they work with the best data they can find, take all their years of experience and expertise, and analyze it, and come up with what we had seen before. What the possibility would be, however, is where all these circles are, is to look at the geomagnetically induced current as it's measured in about 100 or more places across the country. That would be something appropriate to use, I would think. But we don't see evidence of that being used. It, I would guess, would cause us to have a higher standard.

But that's what's available. And that's what some folks have done, including Tom and John Kappenman from Maine. And then, we have just two more slides. The next one—Oh, they're out of the order I thought they were in. Okay, this—And I hope you can see this well enough. There are two blue lines on—two lines on this graph. The one on the left, it shows the modeling done by these independent experts and what they came up with.

What they came up with was actually the red line. What the NERC profile, consistent with the one before, is the blue line. You will see the circle there represents a big difference between what NERC, for instance, would suggest that we worry about, and what these independent scientists have done. That's their modeling. That's the quality of the modeling that they do.

On the right, there is another graph that shows actual incidents. And again, the blue line is NERC's, which shows a much more modest effect of the solar storms on the grid. This is what the actual measured data shows. It's very similar to what you see on the left. And you can see that the peaks are more extreme. So I just wanted to

154

show you, we've got experts that do very high-quality modeling that works out to be quite consistent with the actual measured data because of the care that they give it and what NERC shows us is a much lower standard.

This is simply a conclusion that they had in one of the studies with Kappenman and Birnbach. And I'll just read it to you because I'm not sure you can see it from where you sit. In conclusion, the NERC standard has been defectively drafted because the standard drafting team has chosen to use data from outside the United States and which excludes important storm events to develop its models instead of better and more complete data from within the United States, or over not disclosed—or over more important storm events.

GIC data, in particular, is in the possession of electric utilities and EPRI, but not disclosed or utilized by NERC for standard setting and independent scientific study. The resulting NERC models are systematically biased toward a geomagnetic storm threat that is far lower than has been actually observed and could have the effect of exempting U.S. electric utilities taking appropriate and prudent mitigation actions against geomagnetic threats.

And that's kind of it in a nutshell. And essentially, the profile, the NERC profile is two to four times lower than what's been observed. And, in addition to that, there is usually built in a safety factor of two-to-four times that. So it's way off from where it really should be for major infrastructure that we expect to have a long life. I just wanted to point that out to you. Thank you.

[applause]

TOM POPIK: So Andrea Boland really did an excellent job explaining what could be technically complex. I think there's a couple takeaways. She has an extraordinary talent for distilling something which is complicated into language which regular people can understand and appreciate. That would be Number One. Number Two, this situation with NERC understating the solar storm threat is so egregious that everybody in this society really should be able to get this.

And with that, I'll introduce Ambassador Hank Cooper, who has some really exciting things to talk about in regards to initiatives in his home state of South Carolina.

AMBASSADOR HANK COOPER: Thanks Tom. I'm not—I asked Tom how much time I had, and he said I had five minutes. And there are two or three things I want to say before getting to the one chart I have on South Carolina. First of all, I'm sorry I wasn't here to hear Senator Johnson this morning. But I want to say that, based on conversations that another colleague and I had with him a month or so ago, I think he's going to have a profound impact on Homeland Security matters. And I look forward to that. And on missile defense and EMP issues, in particular, as they apply to the homeland defense, Homeland Security agenda.

The second thing I guess I want to say is that, in case you don't know who I am and what I've done, two things. First of all, about 35 years ago, I oversaw most of the DoD hardening programs for electromagnetic pulse. And so I know what's involved in that matter and how difficult it is to actually maintain systems that are sufficiently hard, that we had confidence that they could be used if the Soviets ever attacked us and the President wanted to retaliate.

The problems were so difficult, for example, with the communication systems, that we had no confidence in AT&T and our normal communications systems to work during those. So we had the programs that built redundant, separate pathways for the President and the Commander of Strategic Air Command to communicate with the bombers and with others. And that's not a well-known fact. One of the individuals that you missed because he's had a heart condition, is Bron Cicotas. And he oversaw the development of that program called GWEN (Ground Wave Emergency Network). So they're real experts that—sorry—that are parts of our community.

The third thing is that I, along the way, was a director of the Strategic Defense Initiative. So I know something about missile defense. Actually, I started working missile defense issues at Bell Labs in the early 1960s. So I am very familiar with the technology work and what it can do.

The fourth thing is that my impression of this community's problem is that we face an existential threat from ballistic missile attack that can produce electromagnetic pulse. Most people want to ignore it. One of the most important things I did here while being here the last two days was Trent Franks yesterday morning, when he observed that he recently read an important report and said the Iranian—some authorities in Iran, at least, had acknowledged EMP as a part of their planning, and how they would employ it.

Put that together with the 1998 evaluation of the threat, and during that time period, the experience that was observed, where Iran launched ballistic missiles off of vessels in the Caspian Sea.

We have no defense against missiles launched from off our coast today. Put those two things together. And in particular, from the Gulf of Mexico, from the south, we have no defense. Associated with that is, we have no defense against satellites that both North Korea and Iran have launched over the South Pole. People are debating in this town whether to build a site on the East Coast to defend against Iranian missiles coming over the North Pole toward us.

But we're completely vulnerable to what they've already demonstrated as a capability of launching a satellite which has a payload capability sufficient to deliver a nuclear weapon 100 or so miles above the United States. You don't have to design anything to reenter the atmosphere. You just set it off while it's up there, and it turns the lights off.

And I have no confidence that the electric power grid is, today, survivable against that kind of a threat. Secondly, I don't have any confidence that anybody is going to harden the grid in the near term to deal with that issue. So I believe the only real alternative in dealing with the issue is missile defense.

Now, if you do EMP hardening against that threat, as a byproduct, you get hardening against the solar threat. If you do only the hardening against the solar threat, as George Baker—I don't know whether he's talked or he's going to talk about it—will tell you, you don't solve the problem for the nuclear EMP. So whatever the politics are in discussing this issue, you should understand that if you work the missile problem, you also work, at least, the solar problem.

Not unrelated to that is the fact that if Iran decides it's going to attack the Great Satan, as they put it, they're not going to attack us only with that. You can fully expect that they'll be kinetic attacks like at Metcalf. And the other thing that Frank has talked about was, they understand the electric power grid is a weakness that we have. So there's no reason to believe they won't target it, both with kinetic attacks, with cyber attacks, what order that's up to them. They've thought these issues through, you can bet. They're technically sound people.

So all of that put together says that we'd better wake up America. Now, I came to the bottom line about two years ago that Washington simply was not going to work this problem. I am completely frustrated by watching what goes on, in the NERC/FERC and all of that mess is just part of the problem.

And I became persuaded that, unless we get people at the grassroots—and by that, I mean the local, state folks involved in this issue, we'll never deal with it. And personally, I believe the National Guard is the framework by which, if we invest our time correctly, we can pull things together in a way that really affects all of America. And I just want to—if you show me the one chart—One reason why I believe that's so, and the role that's out there South Carolina can play, there are two individuals who came to our—Can I have the next chart please? You don't have to read it. You can read it later.

We had a couple of Guardsmen who came to the conference the last couple of days because we're planning to do an exercise down there. And I expect to get some of the key people in this room to come and explain these issues to the folks at home. They don't understand, I suspect, like many people around the country. But it doesn't take long to explain to them that they have serious problems that they need to deal with. And the Guard can play an important role.

How many of you know that the missile defense systems in this country are operated by the National Guard? Homeland defense in Alaska and California, the National Guard is responsible for the operation of those systems. In South Carolina, we have a unit called the 263rd Air and Missile Defense Command in Anderson. It is a deputy, the Commander is a Two-Star Guardsman. He's also the Deputy Commander of First Air Force located at Tyndall Air Force Base in Florida. That command has a responsibility of the Air Defense of the entire continental United States, the Dominican Republic and Puerto Rico.

So they're a deputy to them. They're also a deputy to First Army in Sam Houston, Fort Sam Houston in Texas, which also is a—reports directly to the Commander of NORTHCOM. So networked together the folks in South Carolina are tied in to the defense of the entire country. And I believe, if we can get our act together down there, we can begin to be a bit of glue to tie all of the institutions together, to deal at least with the ballistic missile part of this threat.

One of the reasons I'm interested in the Tyndall connection is that one of the smartest things I believe the country can do, and we're not doing, how many know what Aegis Ashore is? Does anybody know? Okay. We're deploying two sites in Europe to defend Europe against missiles launched out of Iran. One is in Romania, will be operational within a few months. One will be in Poland, operational in 2018, I believe it is. And if we can afford to put sites like that over there to defend Europeans against Iran, it seems to me we could put them at military bases around the Gulf of Mexico to defend the U.S. people against threats from the south.

And so I'm looking forward to going with the Commander, with the 263rd, down to talk to—I think it's General Estes who's a Three-Star about this issue. And, in particular, engaging in a serious way the deployment of a site down there to begin this kind of an active defense. There are a lot of other things that can go into this. And I look forward. I'll report back to you, hopefully with some progress, the next time we have a conference like this. Thank you.

[applause]

CHUCK MANTO: We have at least one question from the audience. Again, rise and state your name.

BOB KOMANSKI: Bob Komanski. I understand that Russia has naval elements in Cuba. Are you talking about our southern borders from that area? Was that a scenario?

AMBASSADOR COOPER: Yes. If I had more time, I probably would have mentioned the fact that it's not so much Russian, North Korea moved a ship out of—where was it? It was in Cuba, I believe, through the Panama

Canal. Was stopped by the Panamanians last May or so. And it was carrying two SA-2s, if you are familiar with that. Those are nuclear-capable missiles that the Russians used during the Cold War. They didn't have nuclear weapons onboard, but they were quite capable of carrying a weapon that could have produced an EMP effect. And they were stopped after—I think after they transited the Canal. So this is a serious threat. It's not just hypothetical.

MICHAEL HOEHN: Michael Hoehn, Alliance for Vigilance. I understand that the Iranian Revolutionary Guard has been building missile bases in Venezuela. Would that be part of the need for the Aegis land-based defense? And could we use destroyers as an interim defense mechanism?

AMBASSADOR COOPER: Yes. Again, five minutes isn't much time to talk about all this. Venezuela and Iran have been working together on this kind of a threat. And if the President chose to put Aegis ships in the Gulf, he could. Of course he directs the Navy. He's the Commander in Chief. But they don't normally go there, which is why I want them on military bases.

If you talk to the Navy about stationing their ships, basically as—what do you call them—just tied to the coastal areas, they get very annoyed by that kind of instruction. Of course they'll—they care about the defense of the American people, and in extreme instances they do that. But they don't want a mission of protecting our shores, per se.

On the East Coast, last year, 2013, we checked the statistics. And, on any given day, at random, there were between four and six Aegis Ballistic Missile Defense (BMD) ships along the eastern seaboard or in our ports. So we de facto have the capability, if we just simply train our crews to operate ships in a way that could provide the Eastern Seaboard a defense, and the same for the Pacific Seabaord. But they don't operate in the Gulf of Mexico. And frankly, I would not advise anybody being a great advocate of having them station themselves in a picket ship role in the Gulf of Mexico. They wouldn't—They wouldn't go for that. The Navy would kill it.

CHUCK MANTO: Any other questions? Okay. Yes, we have a question. Again, identify yourself please.

YVETTE: My name is Yvette. I have a question. I am very happy to be here, but I don't have the background many people have in here. And they are scientists, biologists, can be microbiology study. Here is my question. In the past, I know that Cuba was a strong—having a strong relation with Russia. So is there any intelligent information that we are aware is Cuba is having a role in a missile project to attack to us? I wonder if anyone know anything about it, because Cuba is kind of too close to us. And in the past, they were a strong ally to Russia.

AMBASSADOR COOPER: I don't really, in my own mind, attach Russian threat particularly to the Cuban connection. Cuba can be very annoying on its own without help from Russia. What worries me more than the Russians is the—it's the Iranian threat. You know, these people, the fanatics there, you call them whatever you wish, I call them Jihadis, they want to kill us. I mean they're very—If you bother to just read what they write, they're repeatedly, not just by-the-by, they want to kill us. And they are happy to commit suicide just to kill a few infidels.

So if they could kill a few million or a few hundred million Americans, what's the big deal of risking their own annihilation just to bring a ship close enough to attack us, if they get, and when they get that capability? That's the threat that worries me more than any other, even more than North Korea. And the North Koreans, in my judgment, already have this capability. The Iranians are always present at the North Korean nuclear tests, their technical people are.

And many people, some really knowledgeable key people, including I think I've been reported that Bill Graham, who co-chaired the EMP Commission, and Johnny Foster, who himself was one of the preeminent nuclear designers, and one of the last ones living, in fact I believe this is the case. So we're living on borrowed time, in my judgment, with respect to this threat.

GEORGE ANDERSON: George Anderson. One question. We have to leave in a second. But do we have enough Aegis ships afloat and still in service? And are they staying in service to do this?

AMBASSADOR COOPER: There are 80-plus ships. I don't know what the exact number is now—there will be 85 within a few months—Aegis cruisers and destroyers deployed around the world. We have, I don't know, several in the Pacific supporting our friends there with respect to the North Korean threat. We have ships among the Seventh Fleet and the Fifth Fleet around in the Mediterranean and Indian Ocean and wherever, Sixth Fleet, I guess it is, in the Med.

But anyway, there are some 30–35 ships. I think it's growing to 35 by 2017 or so. So, and there are a total of something on the order of 70 or 80 capable ships that we can modify in a fairly short period of time for modest investments as compared to what we're spending on a lot of other things.

CHUCK MANTO: Okay, very good. Do I see one other hand? Okay. I'm wondering, Gale, do you have a moment? Or do you got to run out for a flight? Oh you got to go? Okay. We have one question back here. Okay, we'll do one more question. And then I'll tell you what we're doing next in just a moment.

SUMAR CHATERGI: My name is Samar Chatterjee of the SAFE Foundation. Ambassador, you've called yourself Ambassador, but you also have a military and defense background you mentioned. Given that, and you said the federal government military and the President is not doing enough. Now why would you think they wouldn't be doing enough, and especially when you said the Iranians and North Koreans are a threat? I don't think, however much they may be crazy, they're not going to risk their own existence, because they may attack the United States. But retaliation would be pretty bad. I mean most people looking at United States' track record of invading countries, it would be pretty bad. So I don't think they are that bad. And therefore, I don't think the concerns that you have expressed doesn't impress me that much. Now I'm not a Muslim, I am from India and a Hindu. So, given that, if we are not afraid, India is not afraid of these, I don't think United States should be that paranoid, including Cuba.

AMBASSADOR COOPER: Well, I don't consider myself paranoid about this issue. I spent all my adult career worrying about the Soviets and the threats to our country. As I said, you're dealing with people who are quite prepared to commit suicide just to kill a few others. And you're dealing with leaders there. When I say leaders, I mean the Ayatollah. I don't mean the President, who is an apologist for their arms control negotiations.

When they get the capability that they can literally level the playing field by destroying the United States, I believe they'll do it. And I think they're going to—they will do maybe the Israelis first, but they'll come after us. I don't know why you wouldn't believe their doctrine anymore than what the Russians really said.

We deterred the Russians for the Cold War. Some people have serious doubts as to whether the deterrents really worked, even during the Cold War, and whether or not we were just lucky—I mean serious people who study these matters. But when dealing with the people who believe as they believe, and have stated the things they state, and act the way they do, you know, I just don't have confidence in relying the security of my family and my kids on hoping that they really won't attack us, which is what you're saying we should do.

CHUCK MANTO: We have another question in the back. Identify yourself please.

JACK PRESSMAN: Hi. Jack Pressman. I'd like to just add something. I can tell you, firsthand, that the three countries outside of the United States that are actively designing and building EMP protection are the South Koreans, the Israelis, and India. And the Indians view this as a very serious threat. They have very bad neighbors. And they do not view EMP as something secondary. They view it as a primary risk to their country. The new prime minister has put this top on the list. So this whole matter of EMP protection amongst our close allies has raised—become very, very senior issue in their defense protection planning.

AMBASSADOR COOPER: And let me just say, since you ask it specifically, I think some of our leaders are just irresponsible. I'll just say it that frankly. As I said earlier, I've lost any confidence in Washington dealing with this issue, which is why I believe we have to go back to the states and work the issue there, and build it from the bottom up.

Defense of the EMP MIL-STD-188-125—One More EMP Knot to Untie

https://youtu.be/V2v_ESpnwE8

Presented at the Dupont Summit on Friday, December 5, 2014

Dr. George Baker, Professor Emeritus, James Madison University

This presentation presents material that explains the role of MIL-STD-188-125 and related testing of the U.S. DoD establishing its usage to protect defense department infrastructure, especially communications. This counters a challenge to the standard brought earlier in the day by speaker Curtis Birnbach. Dr. Baker led technical oversight of EMP programs for the Defense Nuclear Agency and participated in the development of the standard.

Dr. GEORGE BAKER: Well, I'm going to just say a couple of words. I'm not going to do the overview, but I just—There's one thing I do have to do. And my original talk was going to be some EMP knots to untie. But, after some remarks this morning that shattered the validity of the Faraday cage approach to EMP hardening and the MIL-Standard-188-125 in particular, I felt compelled, actually Chuck asked me to say a few words about that.

And several people have come up to me and said, "What's wrong with (MIL-STD)-188-125?" Just a little bit of background on that standard. The standard was developed—Actually, the whole idea of global hardening came out of the mind of Bill Graham, who is the Chair of the EMP Commission. And he felt that, instead of hardening systems piecemeal, we should put them inside Faraday cages.

And so, when I was at the Defense Nuclear Agency at that time, I assembled a team, and it included the Defense Information Systems Agency, was then called DCA. I had all of the service labs, Army Research Lab, Air Force Research Lab, the Naval Surface Warfare Center. And I had Stanford Research get actually—write the draft standards. And then we had a panel of experts from all the military labs get together and bring this thing out.

And we've used this approach successfully, and actually tested the approach successfully on the Minuteman Missile System, the MX Missile System, the "Discus" Satellite defense satellite (DSCS), the ground terminals, the Millstar System, the TACAMO System, the National Military Command Center, the Pentagon, Air Force One, B-1, Pershing. And those are the—I put this list together very quickly after this morning's presentation.

And so, this is a standard. We know it works. There may be some—and I don't know if Paul Hayes is still here. Yeah. Paul Hayes is familiar with tests with Faraday cage protective systems used against RF weapons that the Faraday cage approach works. There may be some unique circumstances that I need to—I need to get with Curtis Birnbach and understand more fully what the situation was where he did the test and seemed to be able to punch through the Faraday cage.

But that hardening approach was based also on electromagnetic compatibility techniques that have been used for decades by the EMC/EMI community, that Faraday cage approach. So I just wanted to dispel any doubt that the Faraday approach, and specifically the approach advocated by MIL-Standard-188-125, it does work, and has been proven to work.

And I was able to thumb through my computer files and find—this is one input in classified, we tested the Pershing Ground Electronics. We had about 12 copies of it. And we found that the main messages we tested it to high-power microwaves, fast-rise EMP, and EMP. And these were time domain wave forms. And up in the upper left there, no units failed or upset when tested in fully sealed—what they were talking about, fully sealed shield, to high-power microwave, fast rise time, enhanced EMP. That would be the EMP that you would get from an EMP tailored EMP weapon, and the EMP that you'd get from a garden variety nuclear weapon. So that's all I want to say right now.

CHUCK MANTO: Do any of you have any questions of George while he's here on the topic? Because I thought it was a great review. Yes we do have one.

MIKE CARUSO: Just some reinforcing comments. George Baker, obviously, is one of the outstanding individuals in this area. And I've got the pleasure and honor of working with Bill Radasky as well. I've been doing shielded enclosures and EMP hardening for 32 years. I have never seen one fail. I've seen them tested in the most extreme conditions. I've seen some outrageous tests done on them. And I've never seen one fail. So to reinforce Dr. Baker's comments, I find it very hard to believe the information that was presented this morning.

CHUCK MANTO: And we have one other question for you, Dr. Baker. One moment.

Unidentified EMP Counterterrorism Caucus Member from OK: I would add that, in legislation that we introduced, our Counterterrorism Caucus introduced out in Oklahoma, we used MIL-Standard-188-125 as the standard that all public utilities operating within the State of Oklahoma would have to meet.

CHUCK MANTO: Any comments about that, or any other thoughts?

Dr. GEORGE BAKER: I think I'd better go. Just in the interest of time.

CHUCK MANTO: Okay, great. Well thank you very much.

END

ZAP!! ....

.... WHAT HAPPENED???
What's your recovery plan?

Maintain Situational Awareness!
With our next generation **IEMI Detector**
IP Addressable, Improved Sensitivity

DON'T LET UNDETECTED <u>INTENTIONAL</u>
ELECTROMAGNETIC INTERFERENCE
END RUN YOUR SECURITY DEFENSES

emPRIMUS™

1660 South Highway 100, Suite 130 - Minneapolis, MN 55331 - 952.545.2051
US Patents 8,773,107 B2 and 8,860,402 B2

Cameo Presentations (Solutions Update by Sponsors and Others)

https://youtu.be/ZqbOfUkhozM

Presented at the Dupont Summit on Friday, December 5, 2014

Advanced Fusion
Avaya
Cyber Innovation Labs
Distributed Sun
Emprimus
ETS-Lindgren
JCTF
MASS
UET
US DHS

These were quick presentations of various groups offering solutions to EMP threats.

CHUCK MANTO: And why don't we start from this side. Mention either the organization or your company and your name, and then the problem and the solution and how to learn more.

DAVE FRASIER: My name is Dave Frasier. I'm a consulting systems engineer with Avaya. I can be reached at 703-362-2683. The technology I was going to discuss, I thought I had two to three minutes, is Fabric Connect. It solves the problems of resiliency, redundancy, and security. It reduces costs. And its basic premise is making the networks as simple as possible, and thereby lessen the complex infrastructure that people have today.

CHUCK MANTO: And you get 10–15 more seconds if you want. You're very concise. That's great. Good model for the rest of you.

DAVE: A lot of people have a misconception about networkings and why they break. The reason why most networking breaks is because people like me working on them, at two or three o'clock in the morning, have to touch a lot of components. Every time we touch something, the likelihood of breaking it increases. With Fabric Connect, it's a minimum amount of touch. The core, you touch it once, put it once, never have to touch it again.

CHUCK MANTO: Very good. Thank you.

[applause]

[00:01:07]

MARK LINEGAN: Good afternoon. Mark Linegan. What I didn't share with you earlier, NORTHCOM, we consider ourselves the paranoid command. We worry about everything so you don't have to. [laughter] But associated with that, one of the things we do on a regular basis is we Red Team ourselves. And, as good as we think we are, we recognize we're not that good. And it's groups like EMP SIG that are out there thinking the hard—hard problems, and come up with solutions, are very, very helpful.

We've had a lot of doom and gloom yesterday and today. There is some glimmer of hope out there. There are smart people and motivated people trying to come up with solutions. I mentioned the SPIDERS project we're working on, cyber secure smart grid, to really focus on some of our DoD's cannot-fail missions. We're also looking to do the transition to the private sector for similar. And we talked about hospitals, where cannot fail mission is somebody's life. But think of your applications, data centers, and so on. So I invite you to reach out to me via email. The information should be available. Thank you.

[applause]

JACK PRESSMAN: My name is Jack Pressman. I'm the managing director for Cyber Innovation Labs. We have launched a new product, EMP GRID, stands for Global Resilient Infrastructure Development. We have been working very, very closely with Mike Caruso, ETS Lindgren, Bill Radasky, and Metatech, and a firm called Armag.

And I was very happy to hear Dr. Baker's comments that the mill standard works, and that Mike backed it up, because we are deploying the first commercially available EMP-and geomagnetic storm-protected data center and mission critical control facilities. They're 6,000 square foot fully shielded, fully protected facilities. Our first one was in Iron Mountain's facility in Boyer, Pennsylvania. The second one, which is coming online, is in Mount Prospect, Illinois. The third one that's coming out of design is at 375 Pearl in Downtown Manhattan. It's got four—the last two have four levels of protection, not only against full EMP, based on the mill standard, but on a geomagnetic storm basis.

We actually do a power kill and do complete self-generation inside the shielded environment. And that power generation is 30 days without external refueling capability. Thank you very much.

ALAN ROTH: I'm Alan Roth. Thank you. I'm Alan Roth. I'm with Advanced Fusion Systems. I've been working with Curtis Birnbach for 28 years. When he first thought of penetrating through a Faraday shield, it was theoretical. It was based on certain principles of physics that he had come across. And we were having problems penetrating killing RPGs, which are Faraday cages. And so we went to the military with this concept. And they gave us this opportunity to produce an EMP generator that could do this. And it was at Picatinny, as Curtis said. And it took some months to put together. It's $20,000. But the job was done.

And it penetrated right through the wall. And the signals showed the decibel loss that one would expect, which is very little, having penetrated through the wall. Well, that's the one place where this was done. And you may say, "Well, maybe that was some circumstance there that was different from anything else that you've seen." And apparently, that's been the case with you who have worked with Faraday cages over time.

But, the good news is, that we have built testing facilities at our plant, and you saw a photo before that Curtis showed of it. We have two other test cells, too, that are even bigger than that. But this is the one that's going to be operational right away. And we're going to be producing EMP generators in our products, starting next month. We've already started on the parts for them, but the full assembly and all is happening next month.

And we do hope, then, to have a testing of these devices to a point where we will really go at it to not just reproduce what we did at Picatinny, but to do much more than that, and to work with others on it, not just us. People may doubt what we do ourselves in-house. But that doubt may be good, because it'll question us, and we'll be able to satisfy those doubts. So look to the future, and maybe it'll prove Curtis wrong. But maybe it will help everybody else to understand how this was done. Okay, thank you.

[applause]

MIKE CARUSO: Hi. I'm Mike Caruso with ETS Lindgren. And ETS Lindgren is a company that's about 60 years old, primary purpose is to deal passively with RF energy. So we do a lot with electromagnetic compatibility, EM shielding, and electromagnetic test cells anechoic chambers. We built the largest anechoic chamber in the world out at Edwards Air Force Base that holds the B-1 bomber. So a lot of experience with it.

We've got over 10,000 installations to our credit worldwide. We've produced over 200,000 RF filters for facility protection. And we've got about 800 employees worldwide. So been around for a while. I head up the HEMP IEMI section of the company. And, if you want to know more about what we do, I'd encourage you to go to EMPauthority.com. And it's a website specifically dedicated to our EMP program. Thank you very much.

[applause]

STEVE COHEN: I'm Steve Cohen. I'm the Vice-Chairman of the EMP SIG. I'm the Vice-President of the Denver Chapter of InfraGard as well. And I'm also the National Coordinator of the JCTF, which means Joint Communications Task Force. JCTF was developed in the city and county of Denver by us, over a period of the last eight years, to address interoperable communications, as well as emergency communications.

When Denver decided it was something that they didn't want to continue to deal with, we decided to take the concept nationally. So we've rolled out the concept to the EMP SIG, to InfraGard, and to the FBI. And it's in the process of developing systems now, organizational charts, and the various elements that are required, just like starting a business.

The JCTF is developing relationships internally within a number of levels of the FBI, inside DHS, and a number of other organizations for the purpose of trying to come up with a cogent plan to have interoperable communications where now the federal government is, in part, siloed in various areas and duplicative in other areas. And our intervention, our consultative capabilities, and so forth, and most importantly, our ability to act tactically, as opposed to just being informational, is one of the primary rationale for the development of the JCTF. And thank you for your attention.

[applause]

ROSS MERLIN: Hi, Ross Merlin, Department of Homeland Security SHARES Program. SHARES is the Shared Resources HF Radio Program. It's been operating since 1988 as a way of coordinating surviving HF long-range communications assets of the federal government, so that no matter what messages could get through. We've obviously been successful since the world as we know it still exists. So it's time to expand the program. Whether your critical infrastructure, key resources, state government, you might have a need for reliable infrastructure, independent communications. And that would be HF radio.

Our office is expanding the program so that, no matter whether you're licensed in the business industrial pool of frequencies, the state government pool of frequencies, or the federal government pool of frequencies, we can provide an interoperability path. The radio regulations seem to be designed to keep those groups apart. Our program is expanding to provide a path to bring them together. So, as we heard before, you may have communications to meet your needs, but your communications need power. Power needs fuel. And both of them need water. And they'll need transportation systems.

And at present, there is no way for a communications asset in one world to reach out to the others. So, if you need help in finding out about licensing HF radio, either for your business or for government or for federal or how to bring it all together with shares, we are available to help you. You can either contact Kevin Briggs, whose contact information will be in the program, or you can contact me, Ross.Merlin@hq.dhs.gov. Thank you.

[applause]

CHUCK MANTO: Thank you. Now you know why I said this is like speed dating, right? Thank you very much. And yes, one more.

DAVE FRASIER: Yeah. This is Dave Frasier, again, with Avaya. You can also look at Avaya website, Avaya. com, in particular look at stealth networkings, a concept called Dark Horse Networkings, devised by a man named Ed Koehler who sits on the IEEE IETF Committee. It's PCI *(peripheral component interface)* compliant. It creates a network environment that is totally impenetrable. Thank you.

CHUCK MANTO: Great. You know, that's an important issue. We're talking about reliable, self-healing, local grids that complement the larger grids. And it's great to be able to do the same thing with networks and have the self-healing local networks that can be islanded just as the others.

Next Steps for EMP SIG Working Groups and Concluding Remarks

Presented at the Dupont Summit on Friday, December 5, 2014

https://www.youtube.com/watch?v=QI8DKNFublE&feature=youtu.be

Mr. Chuck Manto
Mr. Arnold Kishi (HY IMA Chair)
Ambassador Henry Cooper
Mr. Bill Harris
Ms. Andrea Boland

This segment provided an overview of next steps with emerging regional EMP SIGs and the workshops and table top exercises expected over the next year. It also provided a reminder of the EMP SIG meeting in April at the Space Weather Workshop (April 15) in Boulder, CO. Comments were also provided by Mr. Arnold Kishi describing emergency management efforts in Hawaii, Ambassador Henry Cooper describing vulnerabilities of spent fuel cooling ponds at nuclear power plants, Mr. Manto discussing role of VOAD and faith-based organizations in disaster planning and constitutional issues regarding metadata security requirements and privacy issues, Bill Harris commenting on constitutional and regulatory issues, the delegation from Charlotte, NC providing input via webcast, Andrea Boland discussing state level activities around the country... and others. Mention was also made of conference publications available in print and Kindle editions through Westphalia Press and Amazon.

CHUCK MANTO: Now, while we have a few minutes more, I think we have somebody from Hawaii here. Come on up. And then we have—Is there anybody else on the list that I have on the program that I have not included? I think we've got everyone covered. And we're down to our next steps. And let me see. By the way, there was a question somebody had from the website about the (MIL-STD) 188-125. Okay. You answered it for them? Okay, great.

So first of all, what I'd like to do, in the next few minutes, is just open up, for just quick ideas, some of the next steps you think you would like us to take. I'll just mention what a few of them are. Then we'll open it up for questions. We're just going to still conclude by 5:30.

As you know, we're going to take the material we've developed for yesterday's tabletop exercise. We're going to analyze it. We're going to write it up. We're going to gather more information from the collaborators you've heard from the last two days. And we're going to package it and make it available to Northern Command, to DHS, to the National Governors Association, the National Association of Counties, with the expectation that they'll also help gather information on an ongoing basis and make this a joint continued improvement or continual improvement program.

And, as you heard, New York State is going to try to do an exercise coming maybe sometime in March, still under development. You may have some things like that in your local region you'd like to mention as well. And I think in April or May, we'll have a space weather conference in Boulder, CO.

So I would like you to give a few moments of thought of ideas that you might have. But before we do, since Arnold Kishi is from Hawaii, and he spoke briefly last year, maybe he has a thought or two about InfraGard and the resilience of Hawaii and insights as an advisor to DHS.

ARNOLD KISHI: Thanks Chuck. So good program, as usual. So thank you very much. I just thought I would add some perspective on some of the issues that we face in Hawaii. I run the InfraGard program there. And I'm also, as Chuck said, advise the Homeland Security advisor and his service policy advisor to the governor.

What's sobering in listening to the conversation here today is, Hawaii, when it prepares for disasters, we tell all our residents, seven days of wares and food and everything. So it's not three days like you hear here. In Hawaii, it's seven days. So just imagine a week you have to prepare. So, when you go and visit Hawaii, you'll get a message pretty regularly that, if there is a disaster, you're on your own for seven days. And that's because of geographic isolation, as most of you know.

Now, in listening to the long discussions about electrical power systems and so forth, obviously Hawaii is not part of the grid. We're pretty much on our own. And Hawaii, as most of you know, has the highest per kilowatt average, 35 cents per kilowatt, compared to about 11 cents nationally. And that's primarily because everything is fossil fuel up to this point, all petroleum based.

And this has been an issue for a long time, as to how you get industry there to look, for example, at issues of EMP when they're just dealing with the economics of generating power versus investing in the infrastructure to make it more resilient. Well the big news this week, as it was coming up and I left Hawaii, was that the electric utility, which is privately owned, is merging with Next Era, which some of you know is a major electrical producer on the Eastern United States. They have nuclear plants. They have been raised a lot of new technologies in generating and distributing electricity.

So one of the policy issues the state faces is, what kind of limitations or what kind of requirements to put on the acquiring company or the new partnership? And the basis of this merger is because the other partner has a deep pocket, in terms of being able to put in new infrastructure and adopt different technologies that Hawaii has not looked at before. So we see this, at least in the context of the discussion this week here, as an opportunity, when they build that out, to make it much more resilient to consider some of the EMP issues that they could do when they build the foundation, versus many other places, where you have to retrofit those solutions. So that's kind of the context of the opportunities and the challenge we're facing, as related to EMP.

CHUCK MANTO: Did I hear a rumor that said that the company was going to give you a free nuclear power plant if you allow them to make a nuclear waste dump on one of the islands?

ARNOLD KISHI: Not those islands. Perhaps further out in the Pacific.

CHUCK MANTO: Hold on one second. We have a question.

BILL HARRIS: Well, it's partly a statement. The view graph of mine that wasn't put up had one—The heading was, "Can states require, as a condition of utility mergers, that there be a program for higher reliability in the state?" So David Bardin who talked is trying to get Exelon, that's proposing to take over Pepco here, to have a reliability program upgrade. So you ought to consider that. The states, remember, have this authority for reliability. It's something you might consider as part of the merger condition process.

ARNOLD KISHI: That's a good suggestion. I think the new governor who took office on Monday has said that they will be reviewing the conditions on which this merger will occur. And one of the issues is what you just described. And, just as an offshoot, Comcast, which is acquiring Time Warner, and Time Warner is the only provider in Hawaii, public utilities commission is also looking at those kinds of reliability issues for Internet service, for example.

JENNIFER BREZOVIC: Okay, ladies and gentlemen, we have a message from Torry Crass from the Charlotte Group is wrapping up. "Again, excellent conference. The current direction for the group here, InfraGard specifically, is that we will be meeting again soon to discuss local efforts in the Charlotte arena, to engage emergency management, personnel, and other administrative entities, and work with them to raise awareness. Thank you all for the great presentations and information up in DC. We're looking forward to continuing these discussions and helping to move this forward as we can."

CHUCK MANTO: Let's have a round of applause for North Carolina. [applause] Thank you, North Carolina and Gary Gardner. Yes, Ambassador Cooper.

AMBASSADOR HANK COOPER: Just one minute. You triggered one thing I meant to say about, when you mentioned the nuclear reactor power and that sort of thing in Hawaii. South Carolina has five reactors. And they're talking about building two more. There's two—There's good news and bad news with the reactors. In my judgment, that should be reserve power. And if our engineers could figure out how to keep them running if there were a major outage of the overall grid that could be the basis for re-instating the grid. If the engineers don't figure out how to do that, then they'll lose cooling water and pretty soon we'll have 100 Fukushimas around the country.

And so, this is a challenge I'd like to leave to the engineers in this group, figure out how you don't have to scram nuclear reactors if the rest of the grid goes down. Thank you.

CHUCK MANTO: Very good. You can come forward if you've got a quick question. And you're going to be the last one, because I've got about three minutes of wrap-up to do. So you've got 30 seconds to ask, maybe—.

UNIDENTIFIED AUDIENCE MEMBER: I was going to make a statement about the educational outreach efforts. We're working on putting together a media team to document exercises based on the tabletop exercises that we did yesterday, working with EMP SIG groups around the country to document real life exercises with real life people, real life solutions, and use that as an educational outreach for both solutions and for driving awareness and building a support for InfraGard and EMP SIG around the country.

CHUCK MANTO: Thank you very much. Now, that's—I'd like to just mention a couple of items before we close, just to stimulate more thought as to how we engage not only each other, but people we know. One is, how many know the Volunteer Organizations Active in Disaster? Raise your hands if you know that group. Probably just a few, okay.

One of the things we think about is a societal consequences of significant collapse of critical infrastructure, is how we work together. And we heard a lot of research earlier this morning talking about how that happens short term, as opposed to maybe long term. And one of the things we may want to consider doing, and we've had some feedback during the last couple of days, is that we do more to engage the faith community.

I think that's interesting in a couple of ways. Number one, they are the parts of our society who have not just been around for a few hundred years, but a few thousand years. They're the organizations who have had a long memory about things like major disasters, where continent-wide events or near-continent-wide events results in a df of the loss of population. We come up with this phrase "Apocalyptic notions" because of that perspective and long-range memory.

We can learn a lot from those groups. And also, we not only have to create more resilient, critical infrastructure, we need to create more resilient relationships within our families and our communities. And our faith-based organizations and other volunteer organizations can be a way to do that better. And I think one of the things we may want to consider in our discussions is, how do we do a better job reaching out to those organizations and include them in our discussions as technologists or infrastructure professionals?

The second thing I'd like to raise, just from a question for fun discussion. And that is on the cyber-security side. I've had a few discussions with folks who have Constitutional law backgrounds. And I asked a simple question. What's the Amendment to the U.S. Constitution already passed that covers cyber security? And usually, I can't get an answer. They're stumped. And I would think there's at least two. One is the 4th Amendment, which supposedly gives you rights to keep the government from taking your information without your permission— or not without your permission, without a warrant. And then, possibly, the 2nd Amendment, which gives you some rights to defend yourself.

BILL HARRIS: First Amendment.

CHUCK MANTO: First Amendment?

BILL HARRIS: No, the First Amendment is the right to free speech. People—Some want free speech with privacy. And some don't want to go through the barriers others create. So you have a tension between the 1st and the 4th Amendments, basically.

CHUCK MANTO: That's very good. So one the questions I asked, and it would be interesting to do a parallel between the technology and the way we did that in the late 1700s and today. And so, in the day of the 1700s, if I was a citizen, I could watch when I visited Europe somebody building an armada. And I can come back home and say, "Boy, I think we ought to start building a Navy, because they're getting a real doozy of one," right.

Similarly, when I watched the Navy sail over, and there's a bunch of people getting off the ship with rakes, pitchforks or muskets or whatever, and then they look like they're in military formation, I might want to call the National Guard or somebody and say, "Hey, there's somebody coming off a boat, and they don't look very happy." So, I have, as a citizen, I may have a right to expect my government to do something to protect me.

Similarly, when the people come down my highway, or my local street, and they look menacing, as opposed to people on a Fourth of July parade, really happy, I might want to be able to call my local militia and protect me. And I have this right to see this with my own eyes, and make some expectations of my government to protect me when they come down the shipping lanes, when they come down the highways, when they come down my streets. And I have the right to look out my window and see if there's a weird guy standing around the bushes, ready to jump out at me, right? And then, maybe lock my door, or maybe I have some kind of right to protect myself with some kind of a weapon or whatever, right?

What is the security version of that? Do I have the right to expect my government to monitor the information shipping lanes, highways and streets and the alleys and the people outside my doorway? Is that called metadata? Do I have the right to demand that the government have that metadata? But do I also, as an individual citizen, have the right to view that metadata? Because I believe I have the right to look out my window and see if there's a prowler outside ready to pounce, so I can lock my door or call the police, right.

Do I have the right to FOIA (The Freedom of Information Act) Google and Facebook and ask them for all those people who are outside my door, ready to get in, or not? Maybe I have the same rights to metadata that we're arguing about, whether or not the NSA or the FBI has. Who has the rights to metadata? And do I also have that right? Especially when I know that private companies and bad guys have that information just as readily, or maybe more so, than government agencies we are expecting to watch that metadata for us, to protect ourselves.

I am not—I am only raising questions. I am not providing solutions. But I think the interplay of what do we expect our government to do? What must we take responsibility to do ourselves? What rights do we have to do that? are questions that have not been consistently asked, at least from my purview as a layman, and not a legal scholar.

So thirdly, I just mentioned next steps. I would really want to make certain that we have an opportunity to share information. I'd like to make certain that all of you feel comfortable emailing, calling, making certain you swap information amongst yourselves. What I would like to do is send out an email to all of you, just asking for your feedback. And, as you know, we welcome debate, interaction. I've been invited to take—I have a mobile command center, the only one I know of. It's a civilian mobile command center, independently rated to withstand EMP from up to 140 decibels. And I have been invited to test it at some of these emerging new facilities. I've already had it independently tested by folks like SARA, INC.

And so anyway, it would be sort of fun to watch all of us who are technologists interact with each other, let people who come up with great new ideas for weapons to sort of see how well they work, at least in a friendly environment. And then also, see what we can do to engage each other to do other things, to make us more resilient, whether it's energy, food, and everything in between.

Now, I see a hand. Andrea, come on up. You have a question? Because we are about ready to say goodbye. I'd love to have you have the last word.

ANDREA BOLAND: I don't mean to have the last word. I just had neglected to say something I wanted to say before. If anybody is looking to find out what's going on in certain states, I'm trying to keep track of that. There are—There's activity in Oklahoma, Colorado, Florida, Virginia. I had—And I think, you know, there are other places coming along. So, and Utah. So I just wanted to share that with you, in case anybody is looking at their state level.

CHUCK MANTO: Okay, thank you. And just one other comment. Remember, you have a book today, *Mitigating High-Impact Threats*. Last year it was just *High-Impact Threats to Critical Infrastructure*. It should be coming out in Kindle soon. We have an electronic library there that's hyperlinked to well over 80 items. Bill Kaewert did a great job helping us pull that together. We want to make certain that we bring these resources together for you at various affordable rates. So let me tell you, it would take hours—as others who have tried to do this—It takes a lot of time and money to even pull that material together, even if it is in the public domain. And others have contributed stuff privately to make it a very robust electronic library.

So we welcome you to that and encourage your positive constructive suggestions. It's your volunteer help that makes all this possible. So therefore, I would like to give yourselves a round of applause, and our FBI and InfraGard staff today for making this possible.

[APPLAUSE]

Thank you very much. And have a good night.

Risk Communication – Electrical Grid Outage

Ben Sheppard, PhD
Professorial Lecturer, George Washington University
Senior Associate, Institute for Alternative Futures

A sustained near nationwide electrical grid failure and its cascading effects will demand significant warnings and risk communication preparation beyond those developed for more familiar hazards, for example, earthquakes, tornados, or terrorist bombings. Failure to implement effective risk communication and warnings undermines preparedness efforts, exacerbates disasters, and impedes the recovery process. When conducted appropriately, risk communication can significantly increase community resilience, mitigate the cost of disasters, and save lives.

The challenge is how to leverage existing best practices and knowledge to develop customized messaging that can save lives and mitigate economic damage. Food, water, and energy shortages are likely to persist for several months or even years as extremely high-voltage (EHV) transformers are gradually replaced. Effective actionable communications must be prepared prior to an event to push the survival envelope. Readiness for a severe scenario will also better prepare the United States for a less severe and prolonged outage (e.g., a G3 or G4 solar storm, or cyber grid attack on half of the U.S. population).

The demands from a sustained near nationwide electrical grid failure and its cascading effects goes beyond current messaging capabilities from what should be communicated and how. However, there is extensive evidence from our current knowledge base that can guide what can be expected, disseminated, and behaviors to encourage and discourage from the publics.

Two fields are critical to preparing and responding. Warnings on what and how to alert the publics that a major EMP event is about to occur, or for cyber, an attack is in progress. Risk communication on what guidance and messaging to provide to the public in the following days, weeks, and months where possible given limited communication capabilities. Both require actionable information to empower the publics on what they can do and when. Importantly there are multiple publics of various social demographics—not a single public to consider. To assess the communication challenges the article uses the cyber grid attack and EMP solar storm scenario set published for the 2014 *High Impacts to the Electrical Grid: Workshop and Table Exercise*.[1] The no notice cyber grid attack scenario entails hundreds of transformers damaged or destroyed leading to a gradual collapse of the electrical network. By week four around 35% of the United States has power restored. The EMP solar storm scenario envisages a 18–20 hour warning of major flares approaching earth. Not until 10 minutes before the EMP hits the United States will it be possible to confirm that the storm is a G5 event. Seventy percent of the United States is without power with only 50 EHVs transformers working nationwide.

In both scenarios critical facilities including water treatment works have 72 hours of generators. Natural gas flows cease within a week. By week four, the remaining radio stations go off air, there are rumors of social break down, and a barter economy emerges. There is sporadic water and food, the burial of dead becomes a major issue, and outbreak of diseases. Many responders abandon their positions. Full restoration of the grid could take months or years. The death toll from starvation and disease could reach into the millions compounded by severe weather. But many more survive.

The goals for both scenarios is to inform and ready the publics as much as possible when the event occurs and

---

[1] EMP SIG, "High Impacts to the Electrical Grid: Workshop and Table Exercise," December 2014.

before communication systems collapse, push the survival envelope that could make a significant difference in how many survive, and limit the economic damage. There are limitations. The current knowledge base can only provide so much insight into what messages to create and disseminate after the first few weeks given limited communication channels and the gravity of the event. Nevertheless there are findings that can go some way to develop effective messaging although research will be needed to develop a comprehensive approach. The article will be divided into the following. Insights from the warnings and risk communication fields, what actionable communication messages could be provided, and areas that need to be addressed.

Warnings: what and how to alert the publics

A significant amount of research has studied the efficacy of public warnings from building fires to tornado alerts that can inform message requirements. There are a number of key characteristics to consider when developing the EMP and cyber warnings. First, the public is not prone to panic. Panic only tends to occurs when individuals believe there is no escape from a life-threatening situation.[2] Analysis of phone messages from those trapped above the impact zone in the twin towers on 9/11, for example, displayed considerable calmness even when they were aware they may not survive.[3] However, the public can adapt their behaviors to expose themselves and others around them to a greater risk than the original hazard they seek to avoid or mitigate exposure to. This is often referred to adverse avoidance and adaptive behaviors.[4]

A collapse of social order tends to only occur when preexisting social tensions exist. For example, the riots in New York City during the 1977 blackout where years of poverty and high unemployment helped to lay the foundations for the riots and looting that ensued. During the 2003 blackout that struck New York and much of the northeast there was spontaneous humanity with individuals assisting each other and no social breakdown. Shortly after the blackout then Mayor Bloomberg stated New York in 2003 was "a very different place" than in 1977. But the 2003 episode occurred in late afternoon during daylight as opposed to night in 1977. The EMP and cyber scenarios will have blackouts far longer than the hours or days of the 1977 and 2003 incidents.

Second, the warnings need to contain sufficient information to empower individuals on what they can do. Failure to do so may lead to individuals to seek out information from alternative sources that are less accurate, and confusion may result.[5] Third, multiple warnings may be required. A single warning is not sufficient to get people to believe and respond, and poorly crafted warnings and lack of understanding of how the public may respond to a warning will undermine a warning's effectiveness.[6]

Fourth, communicators should be aware of how prior experiences of alerts may impact the reception of an EMP or cyber warning. Those who have experienced warnings that have been later cancelled or received messages without adequate explanation are less likely to believe and/or respond to future warning messages.[7]

---

[2] S. Wessely, "Victimhood and Resilience: The London Attacks – Aftermath," *New England Journal of Medicine* 353 (6) (2005): 548-550.

[3] J. Dwyer, E. Lipton, K. Flynn, J. Glanz, and F. Fessenden, "102 minutes: Last Words at the Trade Center; Fighting to Live as the Towers Dies," *New York Times*, May 26, 2002, p. 3.

[4] B. Sheppard, "Mitigating Terror and Avoidance Behavior through the Risk Perception Matrix to Augment Resilience," *Journal of Homeland Security and Emergency Management* 8 (1) (2011). pp. 1-19.

[5] D.S. Mileti, *Social Media and Public Disaster Warnings*, 2009. Retrieved August 13, 2015, from University of Maryland, College Park, National Consortium for the Study of Terrorism and Responses to Terrorism (START) website: http://www.jeannettesutton.com/uploads/WARNINGS_Social_MediaMileti_Sutton.pdf.

[6] Melissa Janoske, Brooke Liu, and Ben Sheppard, "Setting the Standards: Best Practices Workshop for Training Local Risk Communicators." Report to Human Factors/Behavioral Sciences Division, Science and Technology Directorate, U.S. Department of Homeland Security (College Park, MD: START, 2012).

[7] L. E. Atwood, and A. M. Major, "Exploring the 'cry wolf' Hypothesis," *International Journal of Mass Emergencies and Disasters* 16 (3) (1998): 279-302.

However, publics who previously followed warning messages that proved accurate and effective in mobilizing them to action are more likely to follow future warning or evacuation messages.[8]

The gravity of the EMP or cyber warning may initially be met with disbelief of whether it is genuine. Publics are likely to seek confirmatory signals including what other communication channels are saying (social and traditional media) through to what their peers and family are doing. As with earthquake prediction, key warning challenges include whether the responsibility to inform conflicts with the responsibility of not causing a social disturbance.[9] How can the warnings convey uncertainty for a solar EMP event when warnings three days out may only be 10% accurate? A mid-level storm could turn out to be high level and vice versa. The severity may only be confirmed about 10 minutes before impact. Earthquake warnings may be one area the solar EMP preparedness field could turn to for insights.[10]

Risk Communication: How to survive

Risk communication stemmed from the risk perception research field in the 1970s and 1980s over the need to better communicate risks to the public, namely from environmental pollution and new technologies, for example, nuclear power stations.[11] Following the 9/11 attacks the risk communication field expanded significantly into the area of intentional manmade disasters and with it gained a better understanding of higher dread and lower familiarity events. For example, chemical, biological, and radiological terrorist acts. However the grave nature of a prolonged widespread grid outage and the challenges the cascading effects present arguably goes beyond even preparations for an Improvised Nuclear Device (IND) given the prospect of a near nationwide blackout for months or years in some areas.

Risk communication is best defined as "creating two-way channels, in which recipients are treated like partners, shaping how risks are managed and sharing what is learned about them" (Baruch Fischhoff).[12] Four areas will be important to conduct effective risk communication. Understanding how the publics are likely to perceive a major EMP or cyber risk event; the unique risk characteristics of the event; the importance of trust; and how transparency and uncertainty will play a key role. A good detailed discussion on these areas in an all hazards context can be found in the START risk communication guide prepared for the Department of Homeland Security in 2012.[13]

Publics perceptions

People's risk perceptions of an EMP or cyber event will vary by age, gender, sex, ethnicity, socio-demographics, and proximity to an event.[14] Communicators will need to acknowledge and understand the diverse audience's needs by customizing their core strategic messages for specific population groups. For example, language, how different messages may resonate more with one population type than others.

---

[8] R. Burnside, "Leaving the Big Easy: An Examination of the Hurricane Evacuation Behavior of New Orleans Residents before Hurricane Katrina," *Journal of Public Management & Social Policy* 12 (1) (2006): 49-61.

[9] A. Sol, and Turan H., "The Ethics of Earthquake Prediction," *Science and Engineering Ethics* 10 (2004): 655-666.

[10] A. Sol, and Turan H., "The Ethics of Earthquake Prediction," *Science and Engineering Ethics* 10 (2004): 655-666.

[11] R. Lofstedt, and A. Boholm, "The Study of Risk in the 21st Century," in Ragnar E. Lofstedt and Asa Boholm (eds), *The Earthscan Reader on Risk* (London: Earthscan, 2009), 1-23.

[12] B. Fischhoff, "Psychological Perception of Risk," in *The McGraw-Hill Homeland Security Handbook*, ed. D. Kamien (New York: McGraw Hill, 2006), 463-492.

[13] M. Janoske, B. Liu, and B. Sheppard, "Setting the Standards: Best Practices Workshop for Training Local Risk Communicators." Report to Human Factors/Behavioral Sciences Division, Science and Technology Directorate, U.S. Department of Homeland Security (College Park, MD: START, 2012).

[14] Ibid.

Event risk characteristics
Understanding the unique risk characteristics of the event is essential to build effective messages. It is important to know where the publics are coming from (e.g., information gaps and misperceptions) to know what to communicate. How familiar the publics are with a specific natural or manmade disaster and the degree of dread they have can heighten or reduce the perceived risk.[15] For example, natural weather events like hurricanes and tornadoes have a greater familiarity: events with limited duration and the publics have a reasonable understanding of the dangers they can pose either from direct experience and/or from media coverage and second person accounts. In contrast radiation from a nuclear power leak, for example, is far less familiar and more of a dread risk for the publics. One cannot see, taste, or smell radiation. Its adverse effects on human health and the environment can be delayed and last for several years creating significant uncertainty.

Initially the publics may perceive an extensive nationwide electrical grid failure as another major blackout, and with it view it as a familiar and less dread event. For example, similar to the major electrical outage in the northeast of the United States and Canada in 2003. Generally there will not be an immediate threat to the publics health and well-being compared to say an IND or major earthquake. Hazards from fires caused by failing transformers may be the exception. As days pass the perception will slide more towards the dread and unfamiliar territory as the gravity of the situation becomes clear that the outage may last for months, and what it means for their lives.

Trust
Understanding the degree of trust the publics has in the communicators and the authorities handling the outage will determine how messages are received. Trust between the communicator(s) and the audience(s) is critical. Trust is influenced by perceived competency of leaders and institutions' handling of previous events. A low-trust environment will likely degrade the efficacy of risk messages.[16] The degree of publics trust is likely to vary across the nation according to the population's previous experience in how emergency managers have handled disasters—for good or bad.

Transparency and Uncertainty
Transparency and acknowledging uncertainty will be critical. Messages must be truthful, consistent, and contain actionable information. Conveying the potential gravity of the situation in a calm and effective manner will go a long way to building and sustaining trust for the challenging times ahead. Without which the publics may lose confidence in their communicators and take actions they believe are effective but instead undermines the response efforts.

Actionable communications
The key to success will be to provide actionable risk communication to encourage actions by the publics to limit the risks they face from potential hazards.[17] The manner in which the cyber and EMP risks are presented will either decrease or increase the levels of perceived risk and subsequently influence avoidance and adaptive behaviors—some of which may be detrimental to the well-being of individuals and those around them.[18] Public responses to natural and manmade events regularly illustrate the importance of encouraging behaviors emergency managers desire and discouraging those that may undermine response and recovery.

---

[15] B. Sheppard, *The Psychology of Strategic Terrorism: Public and Government Responses to Attacks* (London: Routledge, 2008).

[16] B. Rogers, R. Amlot, G. J. Rubin, S. Wessely, and K. Krieger, "Mediating the Social and Psychological Impacts of Terrorist Attacks: The Role of Risk Perception and Risk Communication," *International Review of Psychiatry* 19 (3) (2007): 279-288.

[17] D. S. Mileti, L. B. Bourque, M. M. Wood, and M. Kano, "Motivating Public Mitigation and Preparedness for Earthquakes and other Hazards," *Journal of Hazard Mitigation and Risk Assessment* (Spring 2011): 25-31.

[18] C. Keller, M. Siegrist, and H. Gutscher, "The Role of Affect and Availability Heuristics in Risk Communication," *Risk Analysis* 26 (3) (2003): 631-639.

For example, in the months following 9/11 the number of road traffic fatalities increased by as much as 1,595 due to either a greater number of Americans driving out of fear of flying[19] or, as another study suggested, an increase in driving under the influence of alcohol or drugs.[20] While there are various hypothesis to the cause, there was an increase in road fatalities as individuals modified their travel behavior with detrimental impacts.

A strategic communications plan immersed with emergency planning is needed to enable local, state and federal authorities to better anticipate and mitigate adverse avoidance behaviors through becoming aware of:
- Types of adverse public responses they should look out for and possibly expect following an EMP/cyber event.
- Risk communication strategies that can elicit desired behavioral responses and encourage altruistic behaviors.

Behaviors communicators may to seek to modify in the initial hours may include limiting cell phone calls and instead encourage texting and messaging to reduce pressure on communication systems. Encourage the public to stagger essential journeys to avoid a sudden rush to grocery stores, and avoid clogging the roads.

Actionable messages must be developed that take into account the limited availability of communication channels as capabilities degrade once generators at communication stations such as radio and cell phone towers fall silent. At the same time essential utilities of water and gas will gradually degrade along with supplies to grocery stores as the just in time supply chain comes to a crawl.

"Golden hours"

The first few hours following a major EMP or cyber event will be the most important to calmly prepare the publics on how they can endure the likely extensive power outage and its cascading effects including limited federal and state assistance. Depending on the event characteristics, communication channels may have limited capacity during the "golden hours" before the grid collapses and backup generator power is lost at transmitters (e.g., cell, television, and radio towers) and receivers (e.g., cell phone batteries and other home devices). The publics may be more receptive to communications during this time with the event initially feeling more like a regular power cut.

Messages and guidance need to be pre-prepared and ready to go before the "golden hours". Communicators should convey that a paradigm shift is required and to encourage self-sufficiency far beyond what would be expected for other hazards like hurricanes or major winter storms. Research is required on framing the messages and what content to provide. The distribution of a "survival guide" to encourage self-sufficiency may be needed to include the following advice:
- food (storage, preparation, growing your own where possible);
- water (e.g., purification and acquiring rain water);
- sanitation (sewage systems are likely to cease as treatment plants run out of power);
- dead (morgues will be overwhelmed and the dead will need to be appropriately dealt with to limit disease);
- shelter (advice on dealing with extreme temperatures—hot and cold seasons);
- first aid;
- what and when essentials like food supplies and electricity may resume and how future communications will be handled with no power.

---

[19] G. Gigerenzer, "Out of the Frying Pan into the Fire: Behavioural Reactions to Terrorist Attacks," *Risk Analysis* 26 (2) (2006): 347-351.

[20] J. Su, et al., "Driving Under the Influence (of Stress)," *Psychological Science* 20 (1) (2009): 59-65.

The "survival guide" should be disseminated across various electronic platforms that remain active. The publics will need to print out a hard copy where possible given the fluctuating power. Printing presses could be considered to produce extra copies and existing newspaper delivery channels employed to maximize distribution.

Tough decisions will also have to be made to prioritize what populations will receive limited federal and state assistance, and how this should be communicated. Lessons on ethical dilemmas can be drawn from pandemic flu preparedness strategies. Preparedness exercises tackled how decision makers faced with a limited vaccine supply in the face of a severe pandemic could decide to administer smaller vaccine doses to vaccinate a greater number of individuals to increase the number of survivors, but recipients would not know whether they themselves would survive.[21] Similar decisions will need to be taken for an EMP/cyber event. For example, what population groups (e.g., age cohorts) might be best supported to hasten the societal and economic recovery once EHV transformers are back up?

Conclusion

The above is not designed as an extensive roadmap for communications but as an insight into the challenges, and what can be built upon given the current knowledge base. Research is needed to develop ready to go communications and guidance. Initiatives should include workshops to harvest from leading warnings and risk communication practitioners and researchers messaging strategies. Questions to address include: How can we address the uncertainty of conveying ambiguous space weather warnings 20 hours ahead, and could the responsibility to inform conflict with the responsibility of not causing a social disturbance? What warning messages can we provide following a no notice cyberattack that gradually degrades the grid? What guidance should the publics receive on how they can endure several months of a widespread electricity outage with limited external support structures from the lack of food in shops to over stretched federal emergency supplies? What innovative communication options are available once power is gone? What behaviors should we elicit once an event occurs, and mitigate that may undermine recovery? Could this take the form of a "survival guide"?

Current communication practices are arguably limited for the unique challenges of an extensive power outage. Designing messages during the "golden hours" will be too late given the pace systems may degrade and how the perceived dread risk may increase among the public. Coming out strong from the outset with a series of pre-prepared messages and guidance will increase the publics trust in the communicators and their authorities, and ultimately save a significant number of lives as the electrical grid takes months to be restored.

About the author

Ben Sheppard, PhD is a Professorial Lecturer at George Washington University, and a Senior Associate at the Institute for Alternative Futures. He has developed a number of risk communication products for the Department of Homeland Security through START at the University of Maryland, CREATE at the University of Southern California, and Anser. He received his PhD in the terror of terrorism and risk communication strategies at King's College London. Email: ben.sheppard.uk@gmail.com

---

[21] B. Sheppard, "Testing the UK's Response to a Global Flu Outbreak," *RUSI Jane's Homeland Security and Resilience Monitor* (April, 2007): 18-19.

Critical Cyber Vulnerabilities for Robots: Scenarios for 2025
Cyber-Enhanced Well-Being or Artificial Retardation?

August 17, 2015

Trevor Thompson
Ben Sheppard, PhD
Institute for Alternative Futures

Futures Intelligence

The 9/11 Commission Report criticized the U.S. intelligence community for a failure of imagination to anticipate the type of attacks Al Qaeda launched against the U.S. 2001.[1] The futures field offers one such technique to better identify and mitigate attacks on critical infrastructure by pulling together and extrapolating existing data points across the social, political, technical, and environmental drivers. The Institute for Alternative Futures (IAF) has been involved in several important examples of using futures to inform policymakers in national and homeland security. During the height of the Cold War, IAF anticipated that the Soviet Union would recede and Eastern Europe would open up for new market opportunities. During the 1990s, IAF identified that terrorism could become the principle security threat to the U.S. homeland following a devastating attack.

Major cyber-attacks are likely to have unanticipated cascading effects against the U.S. infrastructure as an increasing number of everyday platforms become more interconnected. What new threats do these attacks pose, and how can we mitigate risks before attackers exploit the weaknesses? The accelerating pace of advances in robotics and automation is paving the way for new cyber-threat frontiers. For example, the remote hacking of a Chrysler Jeep in July 2015 to remotely push the car into a ditch is one small illustration of the broader risks we may face in our increasingly connected world.

Robotics and automation are increasingly utilized and relied upon across industries, military organizations, governmental agencies, and several nongovernmental and nonprofit groups. However, cyber-security risk assessments to date have not regularly considered robotics and automation. A scenario approach that merges developments in cyber security, cyber crime, and cyber terrorism with advances in robotics and automation can prove essential in anticipating and preparing for future opportunities and threats at the intersection of both fields. Rigorous scenarios can question and expand assumptions of what is plausible. In doing so, they ensure that strategic planning does not suffer from a "failure of imagination." Scenario planning also helps identify hidden upsides and downsides of strategic options in the threat environment.

For this article, horizon scans were conducted to examine emerging trends in science, technology, economics, environment, policy, and culture that could influence the co-development of robotics, artificial intelligence, automation, cyber security, cyber crime, and cyber terrorism. The results informed the development of forecasts and ultimately of cohesive scenarios. To create the scenarios, IAF's "aspirational futures" method was used to examine how the future of cyber security, robotics, and automation may unfold. This method is specifically designed to stimulate discussion about how to create robust, successful strategies within each and across all scenarios, and to help the public and private sectors create their preferred future.

The scenarios build upon our original scenarios which can be viewed online in *Robotics Business Review* in the March 01, 2014 Guest Editorial.

---

[1] "The 9/11 Commission Report," Final Report of the National Commission on Terrorist Attacks Upon the United States, 339.

Robotics Sector

The robotic sector is projected to grow to over €100 billion by 2020.[2] Robots will also likely become much more ubiquitous by 2025, driven by expansion into new sectors (such as services), as well as by growth in sectors where markets already exist (such as manufacturing). Criminals and terrorists may view robots as an additional attack vector, and might hack into robots to use them as weapons or to collect intelligence for attacks. Cyber events have shown that simple attacks can have devastating consequences, that fake threats can confuse real-time threat assessments, and that attribution can be challenging with spoofing creating fake threats that confuse real-time threat assessments. Attack attribution is often exceedingly difficult. For the robotics sector, cyber-security breaches also pose major financial and reputational risk, potentially undermining robotic market growth in manufacturing, healthcare, and the home, among other sectors. Companies, governments, and designers must therefore prepare for a range of possibilities for the future. This need can best be met via the use of alternative scenarios that explore how the future may develop.

Questions these particular scenarios explore include the following:
- What security risks are emerging from the robot sector?
- What counter measures do we need to start considering today to manage the threats of the future?
- What new employees will companies need to address the cyber threats?

The scenarios are designed to encourage decision makers to ask themselves how they might prepare for and respond to emerging vulnerabilities, threats, and opportunities. The scenarios cover three zones: expectable, desperate, and aspirational. The expectable scenario "Make way for robots" illustrates the view of the most "conventionally expectable future" extrapolated from current robotic trends. The desperation scenario "Cyber insecurity and artificial retardation" presents a set of plausible challenges the robotic sector may encounter from criminal and terrorist groups, and the public backlash that may follow against the robot sector. The aspirational scenario "Robotic and human co-evolution" identifies what a surprising successful counter terrorism and criminal strategy may look like.

Expectable Scenario: Make Way for Robots
The expectable scenario illustrates the view of the "conventionally expectable future" extrapolated from current robotic trends.

Pioneering robotic developments of the 2010s laid the foundation for widespread robotics usage in daily lives over the next two decades. However, cyber-security vulnerabilities in robots persisted, providing openings for terrorists and criminals and sparking significant capital and human investment to ensure the security of new robotic products. By the early 2020s, nearly every city could see robotic babysitters, eldercare robots to compensate for the shortage of human caregivers, and workplace robotic avatars for long-distance meetings. By 2025, it will be hard to see any activity where robots are not part of daily life.

In November 2016 international car manufacturer GBS GBS *(GROUP Business Software)* was forced to halt operations for two months following a cyber-attack on its robotic production line. Unknown to GBS, its compromised robotic nodes had been creating unsafe and inferior cars for six weeks. GBS suffered significant financial and reputational damage. Cyber religious hacktivists "#NaturalBrains" claimed responsibility for hacking into the robots, claiming mankind faced a dark future as manufacturers and governments rushed to greater automation.

---

[2] See http://robohub.org/fascinating-projections-100-billion-robotics-industry-by-2020/ and http://www.nytimes.com/2012/04/14/business/global/kuka-german-maker-of-robots-will-expand-in-china.html?_r=0.

The attack did little to hold back technological breakthroughs and applications of robotics. In 2016, three-dimensional printers had started printing robots. In 2019, the New York Police Department (NYPD) rolled out law enforcement crime prevention drones that could anticipate crowd disturbances including releasing tear gas for crowd control and identify suspicious behaviors. Sniffer drones identified drug smugglers and dealers. In April 2019, during major urban protests in New York City, NYPD riot drones were hacked to release tear gas and tasers over a peaceful crowd, causing the peaceful demonstration to descend into violence.

On a more positive note, breakthroughs in neuroscience and robotics paved the way for customizable and flexible mind-controlled robots for the manufacturing and medical environments. For example, mind-controlled exoskeletons enabled paralytics to perform a wider variety of essential tasks. However, in June 2019, #NaturalBrains launched a coordinated attack on 122 robotic exoskeletons used by U.S. military troops and veterans, causing several injuries and 38 deaths as the systems inflicted physical harm on their users.

By the 2020s, open-source cloud robotic platforms were providing the basis for collective intelligence to reside in huge server farms, rather than in the robots themselves, giving roboticists and robots the ability to share innovations and upgrades instantaneously. Cyclone Bholan that struck Bangladesh with devastating force in July 2022 illustrated the innovation of humanitarian cloud robots that used software developed by an online community to detect cholera and other water-borne diseases. However, the integration of social media and crowd sourcing that helped direct cloud robots clearing roads, moving supplies, and carrying potable water to those most in need and clearing roads also provided a new cyber-security vulnerability. Russian cyber-criminal group Akdov—which also conducted freelance work for the Russian government—brought down the humanitarian cloud robot system for two weeks in order to develop new cloud attack skill sets for the Kremlin to exploit at a time of its choosing.

Cyber-security breaches resulted in some other notable setbacks. In the United States, six million eldercare robots were helping to keep millions out of nursing homes. However, in June 2023, several home-based eldercare robots provided the wrong pills to their patients, leading to 12 hospitalizations. By 2017, car hacking had become the proving ground for "newbie" black hat hackers to demonstrate their abilities. YouTube series were created that showed unoccupied cars being driven out of car parks and driveways to crash into buildings and other objects. In early August 2019, hackers directed robotic cars into truck tankers carrying flammable liquids in Dallas, Texas, and in Milton Keynes, United Kingdom, killing one and injuring four in total. Robotic companies significantly increased their investment in cyber security for robots and cloud robotics thereafter. The expanding robot cyber security and robot repair job market helped to compensate job losses from replacing humans with robots in other workplace areas.

By 2025, neurobots are being used by law enforcement, manufacturers, and emergency services in search and rescue missions. China, however, leads the neuro-robotic manufacturing revolution, having built on Western expertise and research of the previous decade. U.S. President Antonio Suarez, whose 2024 electoral campaign capitalized on public cyber-security anxiety, had imposed stringent robotic manufacturing and licensing regulations. By 2025, cyber-security concerns were ultimately slowing robotics from realizing its full potential for offering societal benefits.

Samples of emerging evidence for....

Humanitarian Cloud Robots

- In 2011, Google announced "cloud robotics"—a free open source operating system for robots. Robot intelligence will be held in a cloud and updates sent directly to connected robots.[1]
- Efforts are underway to combine robots, social media, and artificial intelligence to accelerate disaster relief efforts.[2]

Police tear gas robots

- By April 2015, Indian police had purchased five drones that can use pepper spray during political demonstrations.[3]

Russian Government and Criminal Cooperation

- The cyber stacks on Georgia 2008 and Ukraine 2014 were strongly believed to have been conducted by a combination of government and criminal groups working together to attack the Georgian and Ukrainian government websites and IT.[4]

Robotic Baby-Sitters

- By 2008, there were already an estimated 5.8 million personal service robots—five times more than industrial robots—which included baby-sitter robots.[5] Current baby-sitter robot models functionality include telling jokes, giving quizzes, and using radio-frequency identification chips to track kids.[6]

Eldercare Robots

- Japan is turning to elderly care robots to address a combination of an aging population, falling birthrates, and lack of healthcare workers in nursing homes—a scenario the United States is set to face as baby boomers retire.[7],[8] Japan's Prime Minister Shinzo Abe's government allocated ¥2.39 billion in the fiscal 2013 budget to develop caregiver robots. Baby harp seal robots are already being used in the United States and Japan to provide animal therapy for the elderly and hospital patients.
- In 2013, the University of Salford in Manchester, United Kingdom developed a prototype elderly care robot which can monitor patients, communicate with doctors, and provide basic companionship.[9]

- The International Federation of Robotics anticipates global sales of robots for elderly and handicap

---

[1] M. Ford, "Google's Cloud Robotics Strategy—and How It Could Soon Threaten Jobs," *Huffingdon Post*, March 1, 2012, http://www.huffingtonpost.com/martin-ford/googles-cloud-robotics-st_b_1179203.html.

[2] N. Davies, "Disaster Response Gets Boost from AI, Crowdsourced Data," *Extreme Tech*, June 18, 2015, http://www.extremetech.com/extreme/208180-crowdsourcing-data-for-humanitarian-disaster-response.

[3] K. Knibbs, "Police in India Will Use Weaponized Pepper Spray Drones on Protesters," *Gizmodo*, April 8, 2015, http://gizmodo.com/police-in-india-will-use-weaponized-pepper-spray-drones-1696511132.

[4] HIS Jane's, "West Accuses Russia of Cyber-Warfare," *Jane's Intelligence Review*, December 28, 2014, http://www.janes.com/article/47299/west-accuses-russia-of-cyber-warfare.

[5] B. Keim "I, Nanny: Robot Babysitters Pose Dilemma," *Wired*, December 18, 2008, http://www.wired.com/2008/12/babysittingrobo/.

[6] R. Lynch, "Japanese Firm Softbank to Sell $1900 Robot Babysitter," *Evening Standard*, June 5, 2014, http://www.standard.co.uk/business/business-news/japanese-firm-softbank-to-sell-1900-robot-babysitter-9491219.html.

[7] C. Crisostomo, "Robots: Japan's Future Elderly Care Workers," *VRWorld*, January 22, 2015, http://www.vrworld.com/2015/01/22/robots-japans-future-elderly-care-workers/.

[8] M Iida, "Robot Niche Expands in Senior Care," *Japan Times*, May 10, 2014, http://www.japantimes.co.jp/news/2013/06/19/national/robot-niche-expands-in-senior-care/#.Vc3uGvlViko.

[9] "Salford PhD student develops revolutionary elderly care robot," University of Salford, February 26, 2013, http://www.salford.ac.uk/business-school/about-salford-business-school/salford-business-school-news/salford-phd-student-develops-revolutionary-elderly-care-robot.

assistance will be about 6,400 units in the period of 2013–2016.[10]

- By 2050, 16% of the global population will be over 65 (1.5 billion people).[11]

Neurobots and Mind-Controlled Exoskeletons

- In 2015, a paralyzed patient from the neck down received a brain implant that allowed them to control a robotic arm.[12]
- In 2013, University of Minnesota researchers developed a non-invasive brain-controlled computer interface that can control a flying robot.[13]
- A team at the University of Houston created the "NeuroRex" exoskeleton by modifying a Rex Bionics exoskeleton with an electroencephalography cap.[14]
- Neuroergonomics research combines neuroscience and ergonomics to match technology to the capabilities and limitations of people so that they may work effectively and safely together to inform the design of technologies in the workplace and at home. Neuroergonomics could lead to robots that emulate or are part of human beings.[15]

[10] "World Robotics 2014 Service Robots," IFR International Federation of Robotics, http://www.ifr.org/service-robots/statistics/ (accessed August 15, 2015).

[11] "Global Health and Aging," National Institute for Ageing, https://www.nia.nih.gov/research/publication/global-health-and-aging/humanitys-aging (accessed August 14, 2015).

[12] M. Fox, "Mind-Controlled Robot Arm Gives New Freedom to Paralyzed Man," *NBC News*, May 21, 2015, http://www.nbcnews.com/health/health-news/mind-controlled-robot-arm-gives-new-freedom-paralyzed-man-n362741.

[13] C. Ngak, "Researchers Unveil Mind-controlled Flying Robots," June 5, 2013, http://www.cbsnews.com/news/researchers-unveil-mind-controlled-flying-robots/.

[14] B. Burton, "Next up in Robot Suits for the Paralyzed: Mind Control?" CNET, November 20, 2013, http://www.cnet.com/news/next-up-in-robot-suits-for-the-paralyzed-mind-control/.

[15] "Center of Excellence in Neuroergonomics, Technology, and Cognition," George Mason University, http://centec.gmu.edu/.

Desperation Scenario: Cyber Insecurity and Artificial Retardation
The desperation scenario presents a set of plausible challenges the robotic sector may encounter

Robotics advanced rapidly throughout the 2010s, outpacing societal knowledge of robots, robotics, and artificial intelligence. However, robotics research failed to account for the increasing strength and numbers of hackers. As open-source platforms enabled more people to learn programming, coding, and hacking skills, and as automation increasingly replaced human labor, a movement of hackers against robots arose. By the late 2010s, hackers could easily override industry cyber-security safeguards for robots. Hackers frequently disrupted manufacturing processes and corporate operations, harmed product quality, and stole proprietary information by hacking into industrial robots.

The darkest hour came on September 10, 2018 when a hacking group ISAQ announced it would execute 16,000 Americans and Western Europeans at noon Eastern time the following day. In the early afternoon of September 11, emergency rooms around the United States and Europe reported a surge in heart attack patients. Pacemakers and remote heart monitoring systems previously celebrated as providing effective cardiac care were turned into weapons. ISAQ remotely hacked and disrupted eldercare robots that were connected to pace makers as part of the "Internet of things" for senior citizens. Mass heart attacks were administered at a key stroke as the robots hacked the pacemakers. The day became known as "black heart."

Soon after black heart, a more subtly coordinated attack against IBM's Watson was discovered. Several members of ISAQ and their radical allies around the world had long planned and had finally succeeded in remotely hacking IBM's Watson, which had become a staple of healthcare for many public hospitals and private healthcare facilities. Hackers had managed to "confuse" Watson's knowledge of appropriate treatments, as well as infect Watson with "digital Alzheimer's." The problem had somehow remained undetected until six young children died after Watson had recommended a very high dose of an unapproved drug for attention deficit hyperactivity disorder (ADHD). After experts analyzed Watson, they determined that Watson may have been compromised for months prior to these parents' discovery contributing to dozens of deaths.

The public was also plagued by the misuse of robots, which hurt the private sector even more. Abuse of drones by hobbyists and law enforcement only aggravated overwhelming and misinformed perceptions of drones from the 2010s. By 2020, some police departments such as the New York and Los Angeles police departments had begun to use drones, or even directly program them, to specifically target minorities. In response, vigilante hacktivist networks frequently disrupted and hacked into police and military drones, and dispersed techniques for how to hack into and disrupt drones. Unfortunately, this also hurt efforts to use drones and robotic technologies for other, more beneficial purposes.

2023 was the turning point. Cyber and hacker terrorists used self-driving cars and drones to transport and set off various chemical weapons and bombs in an attack on Los Angeles and its police department. The consequent large-scale fear of robots resulted in several state bans on drones. This fear eventually evolved into backlash against expansive use of robots in the private sector. People argued that "big corporations" were eliminating jobs for U.S. residents by replacing them with "soulless machines." "Human Touch" emerged as an activist group protesting widespread use of robots in human societies, fiercely opposing private and public sector use of robotic technologies. Human Touch also created a short film in 2025 opposing artificial intelligence research. The film showed a hacker making an artificially intelligent hospital care robot "angry." By the end of the film (which had gone viral) the robot had killed patients, doctors, and children, and had set fire to a patient ward.

In the United States and abroad, the misuse, abuse, and misunderstanding of robots hindered research, development, deployment, and acceptance. In warfare, attack drones were sometimes used to drop chemical and biological weapons on civilians and troops alike. This was commonly referred to as "robotic hell-fire" by human rights activists. Taking advantage of cyber-security shortcomings in robots, skilled hacktivists were sometimes able to turn attack robots against their "owners," and using drones to hack into and control other nearby drones for swarm attacks. Surveillance robots and even robotic exoskeletons were also subject to attacks. In spite of these events, multilateral attempts to prevent the use of attack robots in warfare were thwarted by superpowers and their allies.

As cyber warriors saw the potential for disrupting robots on the battlefield, radical vigilante hacktivists perceived a new space of malicious opportunity. An elderly Japanese politician was killed after a radical hacktivist reprogrammed his homecare robot to give him a dangerous mix of pills rather than his prescribed medications. Hacktivist assassinations became more commonplace throughout China, Japan, and Korea, creating regional paranoia of the very robots that many had previously been raised to trust.

By 2025, public outcry against widespread robotics has resulted in huge industrial losses. Cyber security for robotics is notoriously weak, and research is far behind the collective prowess of malicious hackers. Even the use of robots in corporate or government meetings or for manufacturing processes has been deterred, as an almost tribal movement of humans against robots takes its toll on the private and public sectors.

Samples of emerging evidence for....

Robots Hacking Each Other
A program has been created called Skyjack that can enable a drone to hack into and control other drones flying in the area. Commercially available drones can also be programmed to track individuals using an Android device.[1]

Hackers and Hacktivists Becoming Involved in Robotics
- "Commander X" of Anonymous and the People's Liberation Front (PLF) quit Anonymous and closed down the PLF in 2013, refocusing his efforts "on autonomous robotics and artificial intelligence."[2]

Increasing Potential for Automation to Replace Human Labor
- An Oxford University study showed nearly half of U.S. jobs could be at risk of automation and computerization.[3]

Expansion of Coding, Basic Programming, and Hacking Competence and Learning
- The market for online instruction in programming and Web construction, as well as for iPhone apps that teach has been booming in recent years reflecting a national trend of more people moving toward technical fields. The Computing Research Association identified a 10% increase in the number of students enrolled in computer science degree in 2010—a trend that has continued through 2015.[4]

Hacking of Drones and Autonomous Cars
- In July 2015, journalists at Wired successfully remotely hacked into, took control, and forced a Jeep car in a ditch.[5]
- Car hacking is not new: in 2010, University of Washington and the University of California, San Diego researchers wirelessly penetrated the same systems on a Sedan through Bluetooth bugs, a rogue Android app, the car's OnStar-like cellular connection, and a malicious audio file on a CD in the car's stereo.[6]
- Anonymous hacked into drone maker Alpha Unmanned Systems website, in May 2012, and pointed out the lack of "proper encryption" for manual remote control of unmanned aerial vehicles (UAVs) and video transmission.[7]
- The threat to hacking drones has been proven for some years when in 2012, University of Texas at Austin hacked and remotely controlled drones via inexpensive GPS "spoofing" systems for the Department of Homeland Security (DHS) and the Federal Aviation Administration (FAA). Iran probably used this technique to capture a U.S. drone in 2011.[8]

---

[1] "The Impending Use of Commercial Drones in Stalking, Hacking, and Terrorist Attacks," *Work Place Tablet*, July 27, 2015, http://workplacetablet.com/2015/07/27/the-impending-use-of-commercial-drones-in-hacking-and-terrorist-attacks/.

[2] N. Anderson, "Homeless Hacker Commander X Quits Anonymous, Retreats to Robot Lab," *Wired*, September 4, 2015, http://www.wired.co.uk/news/archive/2013-09/04/commander-x-quits.

[3] D. Rivero, "'47 Percent' Of U.S. Jobs are at Risk Because of Advancing Technologies," *Huffington Post*, March 6, 2015, http://www.huffingtonpost.com/2015/03/06/jobs-risk-technology_n_6817236.html.

[4] J. Worthman, "A Surge in Learning the Language of the Internet," *New York Times*, March 27, 2012, http://www.nytimes.com/2012/03/28/technology/for-an-edge-on-the-internet-computer-code-gains-a-following.html?_r=1.

[5] A. Greenberg, "Hackers Remotely Kill a Jeep on the Highway—With Me in It," *Wired*, July 21, 2015, http://www.wired.com/2015/07/hackers-remotely-kill-jeep-highway/.

[6] J. Markoff, "Researchers Show How a Car's Electronics Can Be Taken Over Remotely," *New York Times*, March 9, 2011, http://www.nytimes.com/2011/03/10/business/10hack.html?_r=0.

[7] "Anonymous Hacks, Defaces Drone Maker," May 27, 2012, http://www.examiner.com/article/anonymous-hacks-defaces-drone-maker.

[8] J. Roberts, "EXCLUSIVE: Drones Vulnerable to Terrorist Hijacking, Researchers Say," June 25, 2012, http://www.foxnews.com/tech/2012/06/25/drones-vulnerable-to-terrorist-hijacking-researchers-say/;%20http://www.bbc.co.uk/news/technology-18643134.

- Back in 2013, MIT Associate Professor of Aeronautics and Astronautics Missy Cummings warned that hackers and terrorists could take over flying robots.[9]

Resistance to Automation and Drones
- The International Longshoremen's Association conducted a rally in New Jersey to protest plans to create an automated port on the Jersey City–Bayonne border that will not require humans. The plans are viewed as a threat to the jobs of longshoremen.[10]
- In 2015, Elon Musk and Stephen Hawking, together with hundreds of artificial intelligence researchers, called for a worldwide ban on so-called autonomous weapons, warning that they could set off a revolution in weaponry comparable to gunpowder and nuclear arms.[11]

The Campaign to Stop Killer Robots, an international collaboration of nongovernmental organizations (NGOs) launched in April 2013 to ban fully autonomous weapons. Some groups were involved in anti-personnel landmines, cluster bombs, and blinding lasers campaigns. A report released in 2015 urges killer robots be banned.[12] In -2012, Human Rights Watch issued the document *Losing Humanity: The Case Against Killer Robots*, arguing that autonomous weapons threaten human rights and could actually increase conflict. Human Rights Watch proposes an international ban on the manufacturing, development, and use of autonomous weapons.

[9] "Why Everyone may have a Personal Air Vehicle," *BBC News*, October 31, 2013, http://www.bbc.com/future/story/20131031-a-flying-car-for-everyone.

[10] S. Strunsky, "Longshoremen Rally against Dock Automation Eliminating N.J. Jobs," NJ.com., January 7, 2012, http://www.nj.com/news/index.ssf/2012/01/longshoremen_rally_against_doc.html.

[11] D. Victor, "Elon Musk and Stephen Hawking Among Hundreds to Urge Ban on Military Robots," *New York Times*, July 27, 2015, http://www.nytimes.com/2015/07/28/technology/elon-musk-and-stephen-hawking-among-hundreds-to-urge-ban-on-military-robots.html?smid=fb-nytimes&smtyp=cur&assetType=nyt_now&_r=1.

[12] O. Bowcott, "UN Urged to Ban 'Killer Robots' Before They Can be Developed," *The Guardian*, April 9, 2015, http://www.theguardian.com/science/2015/apr/09/un-urged-to-ban-killer-robots-before-they-can-be-developed.

Aspirational Scenario: Robotic and Human Co-Evolution

The aspirational scenario identifies what surprising success can look like to address cyber-criminal and terrorist acts in the robot environment.

The "One Robot Per Family by 2030"—goal announced by the philanthropic group OneVision in 2020—received tremendous support from global leaders and netizen advocates. "One Robot Per Family" aimed to improve life for every family within a decade. The vision sought to use robots to dramatically improve conditions for children and families in developing countries, while simultaneously addressing the needs and burdens of aging populations and youth in more developed countries. This goal, along with several research breakthroughs, revolutionized robot usage and design in the manufacturing sector. However, serious security concerns had to be addressed to protect society from cyber criminals and terrorists. The cyber-security approach of the early 2010s was failing and unfit for the task of One Robot Per Family by 2030.

As demand for robots skyrocketed, the robotics industry redefined cyber security beyond the twentieth century constraints of defending against cyber-attacks and ensuring the safety and security of robots and their users. Cyber security now implies the securing of benefits that computers can offer to humans. This was labeled "Cyber-Enhanced Well-being." Cyber-Enhanced Well-being was based on the ideas expressed by Marshall McLuhan and Father John Culkin: "We shape our tools, and thereafter our tools shape us." Humans had initiated the evolution of robots and robotic technologies, and now increasingly came to accept and benefit from smart use of robotics. In this sense, robots and humans began to co-evolve and mutually improve each other. Cyber-Enhanced Well-being advanced the notion of cyber security to ensure that security, equity, and well-being were enhanced via programming and software schemes integrated into our daily lives. IBM's smart cities proved a precursor to this, as cyber security served as a foundation for Cyber-Enhanced Well-being that helped design cities, city functions, and city processes in which humans do not simply live, but thrive.

By the early 2020s, the complex demands of Cyber-Enhanced Well-being were driving a new interdisciplinary robotics jobs sector. The meaningful and high-paying sector was comprised of ethicists, psychologists, roboticists, sociologists, teachers, programmers, engineers, and neuroscientists. It was established to design and maintain the "new age" robots, thus creating significant employment opportunities. Robotic associations and the robotic industry had years earlier successfully encouraged and co-designed new university degree schemes in order to help the emerging workforce meet its predicted specialist demands. By 2023, the creation of neurobots and neurobot human–machine interfaces—the results of advances in neuroergonomics, neuroscience, biomonitoring, and cognitive computing—had allowed manufacturing to increase productivity and product complexity, while increasing the longevity of the workforce as retirement ages increased in more developed nation-states.

In the late 2010s, the robotic manufacturing sector was at the forefront of crafting policies and regulations to enable Cyber-Enhanced Well-being. In 2021, robotic "rights" were established for the manufacturing and home sectors, as increasingly advanced robots became central to family units. Killing, hacking, or "infecting" a robot were often met with prison terms focused on reeducating violators. Self-learning hive minds, robotic "crowdsourced" knowledge and learning, and a growing collective "awareness" among robots required stringent protective measures to prevent robotic security breaches. The Robotic Warfare Treaty of 2022 outlawed attack robots, likening them to chemical and biological weapons.

By the mid-2020s, artificial intelligence developments and big data on optimal human interactions were helping robots to assure that human interactions were good, healthy, and uplifting—and rooted out cyber-attacks at their source. The foresight of Cyber-Enhanced Well-being ensured the conscientious care for robots by humans, ethical use of robots by humans, and the security of robots as they cared for and interacted with humans. By 2025, Cyber-Enhanced Well-being enables humans to improve themselves, their behaviors, and their interactions.

Samples of emerging evidence for....

Robots in Every Household

- South Korea's Ministry of Information and Communication has projected that networked robots will be in every South Korean household between 2015 and 2020. The Ministry is tasking over 1,000 researchers and over 30 companies to make robots more ubiquitous in South Korean society. Networked robots are expected to undertake tasks such as patrolling public areas, teaching children English, searching for intruders, guiding customers at post offices, and sending images to monitoring centers.[1]

Increased Demand for Robotics

- Markets and Markets forecast an annual 21.5% growth between 2014 and 2020 in the service robotics sector, creating a $19 billion industry by 2020.[2] The International Federation of Robotics projects that 23.9 million service robots for personal use will be sold between 2014 and 2017.[3]

Advocacy of Robot Rights

- The American Society for the Prevention of Cruelty to Robots (ASPCR) argues that, when robots and "Created Intelligences" become self-aware and intelligent, they should have the rights to "Existence, Independence, and Pursuit of Greater Cognition."[4]
- South Korea's Ministry of Commerce, Energy, and Industry commissioned a team of futurists and a science fiction writer to draft a Robot Ethics Charter that would prevent the abuse of androids while programming ethical standards into robots. The Ministry commissioned this charter in anticipation of intelligent service robots becoming routine parts of daily life. The charter also deals with protection of data that robots acquire.[5]

---

[1] N. Onishi, "In a Wired South Korea, Robots Will Feel Right at Home," *New York Times*, April 2, 2006, http://www.nytimes.com/2006/04/02/world/asia/02robot.html.

[2] "Service Robotics Market (Professional and Personal), by Application (Defense, Agriculture, Medical, Domestic & Entertainment), & by Geography—Analysis Forecast (2014–2020)," *Markets and Markets*, September 2014, http://www.marketsandmarkets.com/Market-Reports/service-robotics-market-681.html.

[3] "World Robotics 2014 Service Robots," International Federation of Robotics, 2014, http://www.ifr.org/service-robots/statistics/.

[4] See http://www.aspcr.com/index.html.

[5] H.B. Shim, "Establishing a Korean Robot Ethics Charter," *Ministry of Commerce, Industry and Energy, Korea*, April 14, 2007, http://www.roboethics.org/icra2007/contributions/slides/Shim_icra%2007_ppt.pdf.

About IAF

Over the last 35 years, IAF has looked at the future of homeland security, national security, technology, health, and healthcare from multiple perspectives to help companies improve strategy and assure critical decisions made in the context of strategic risk. Based in Alexandria, Virginia, United States, IAF is a 501(c)3 research and education organization founded in 1977 by Clem Bezold, Alvin Toffler, and Jim Dator. IAF and its profit group Alternative Futures Associates develops customized cyber security for robot scenarios; cyber security for robots war game simulations; and cyber decision mapping.

Visit us at www.altfutures.org. You can email us at bsheppard@altfutures.org or tthompson@altfutures.org, or call us at +1 (703) 684 5880.

The original article is part of the "Blackout" issue of November 2014 and can be found here: http://www. domesticpreparedness.com/First_Responder/Emergency_Management/Community_Preparedness_for_ Power-Grid_Failure/

Community Preparedness for Power-Grid Failure

by MARY D. LASKY
Tuesday, November 25, 2014

When emergency managers perform their jobs well, citizens may feel the government will be there to help in times of emergency. For emergencies that last a few days, most people are able to take care of themselves. Many people may not have considered what happens when emergency managers are not able to respond because of the severity of the event. Planning for large disasters is difficult and there is a fear that even discussion could cause panic among citizens. However, when able to think about it ahead of time, organizations and members of the public would know what they can do to prepare for a major disaster, and thus reduce panic.

Loss of the Electric Power Grid: The grid is the generation and distribution of electrical power in interconnected local and region systems across the entire country. It is possible for certain events to cripple or disable parts or all of this interwoven structure. As a nation, the United States thus far has been fortunate to not lose the entire grid.

A cyberattack could disable a local area or an entire region. It is conceivable that a group of coordinated cyberattacks could disable several regions simultaneously. Depending on the level of disruption to industrial controls that also could cause equipment damage, the length of time power would be lost would depend on the length of time before damaged equipment could be replaced. Given enough damage to parts requiring a long time to replace, outages could last from days to months.

In September 2014, a solar coronal mass ejection, or solar storm, came very near to disrupting power. A Lloyds of London report published in 2013 describes that a power outage across the northeastern United States could last more than a year. A manmade electromagnetic pulse (EMP) could happen if terrorists were to explode a nuclear weapon over the continental United States. This would disable the power grid over a wide area by destroying both microelectronic controls and large transformers. Major transmission transformers are not stockpiled and most are manufactured outside the country. Thus, replacing a damaged transformer could take more than a year.

Ideally, the grid would be protected or hardened against these destructive measures, and businesses, universities, and communities would be capable of generating their own emergency power. However, the United States is not yet at this ideal state. Consequently, if the grid is damaged, everyday life would change drastically. If several major locations are involved at the same time, then mutual aid agreements among emergency managers might be difficult to honor and the local situations could become devastating.

Potential Cascading Events: No matter the cause of a grid collapse or failure of parts of the grid, a series of events could follow—a cascade of tragic proportions. Immediately following a power outage, major emergency generators start automatically and could continue running until fuel is exhausted. However, with a large-scale grid failure, multiple infrastructures, on which refueling depends, would eventually fail, including the financial sector, transportation, oil refineries, as well as law and order. Fuel would quickly become scarce. Even natural gas supplies could become depleted when compressors are driven by electrical power. Within one week, or likely sooner, most backup generators will have exhausted their fuel supplies.

Without electricity, both fresh and wastewater treatment plants would fail. People with their own wells could be affected because the wells usually require an electric pump, which could be powered during a power outage by a generator; but eventually the generator will stop without its own fuel source.

Without trucks operating, food distribution would halt. Existing food on shelves would last for a few days at most. Ideally, big grocers would initiate rationing so allocation would be fair and people would not hoard. However, because of just-in-time inventory practices, food is no longer stockpiled in warehouses. The combination of desperate people and just-in-time distribution would quickly exhaust food supplies.

Hospitals and nursing homes would find that, without electricity, water, sanitation, transportation, and other resources, they could not treat patients. Hospitals would shut down. Medical supplies would be in short supply and people would die. People may not have transportation to get to work, so businesses would stop and, with sustained power outages, there may be little work to do.

In addition to the likely consequences discussed above, unpredictable events also may occur. To better prepare for the unexpected, individuals, organizations, and agencies should determine what is controllable and take action now to mitigate the consequences.

Promoting Individual Actions: Emergency managers often promote personal and family preparedness as a three-day supply of water, food, and medications. In reality, three-week or even three-month supplies may be needed. Although individuals and families should be prepared at home, at work, and even in their cars, only a small portion of the population is likely to be adequately prepared when an incident occurs. It is unclear how many could survive for a long period if they were away from their home. As such, emergency managers should inform the public about potential scenarios, so they can be better prepared:

- *Without power, phone systems, and the Internet eventually would stop functioning.* Ways to communicate with family members should be discussed ahead of time. It is important to have a family plan for where to meet and how to communicate without electronic means. With no television or radio, the feeling of isolation and the lack of information would be prevalent. Going to the local fire station might be a place to learn what is happening.

- *Without power, there could be a shortage of water supplies.* Having a generator at home would help with the immediate aftermath of a power failure. Hot-water storage tanks could help in the short run. Rain barrels on gutter downspouts also could provide water in areas with ample rain. However, incredibly, some states have outlawed the use of rain barrels, claiming the rain belongs to the government.

193

In addition, emergency managers could share information with the public about alternative sources of power to assist during long-term power outages. One suggestion is installing solar panels on homes or on poles in backyards. However, since most people who install solar panels do so to cut their power bills, the power generated from the solar panels often goes to the grid. When the grid is not functioning, the solar panels do not provide power to the homes.

This safety mechanism prevents electrical power company crews from being electrocuted by the power coming from the panels while they work to make the grid operational again. To power the home during an outage, homeowners would need to have a switch so they can isolate their solar panels and have the power go to a backup battery. Such switches and batteries, though commonly available, are not commonly used; however, if demand rises, so would the prevalence and affordability of these devices.

Communities Banding Together: Although businesses may be able to function temporarily following a power-grid failure, workers eventually would need to return home to their families. Consequently, businesses including grocery stores and food markets may cease to function.

In urban areas where growing one's own food is not practical, there are some community options that emergency managers could suggest. Victory gardens, where communities garden a plot together, might be a partial solution. Some building owners have the ability to establish gardens on their roofs, which would provide security from renegades. Having an assured source of even a small amount of food could increase survivability. One concern, though, is that well-prepared citizens who do have food would become targets for attack to obtain the food they have. The way to protect against such situations is for neighbors to band together.

Neighborhood groups could consist of a few blocks or an entire county. Volunteer organizations such as "Volunteers Active in Disasters" (VOAD), Boy or Girl Scout groups, faith-based organizations, and book clubs, also could be considered "neighborhoods." Working together and pooling resources may keep everyone alive longer; there is "safety in numbers."

Before a disaster occurs, these neighborhoods should: (a) share ideas and determine what to do in case of a disaster; (b) recognize each other's strength and needs; and (c) discuss major issues, such as the failure of electrical power for an extended period. For example, the public–private partnership organization, Community Emergency Response Network Inc. in Howard County, Maryland, has regular meetings to discuss various emergencies and related steps toward preparedness. Citizen Corps and Community Emergency Response Teams (CERT) operate in a variety of ways within many communities. Relationships built ahead of time will strengthen the entire fabric of society. After an emergency is not the time to share business cards.

To strengthen communities, it is important to build relationships, share information, and know where to turn in times of a disaster. By helping neighborhoods expand their relationship circles, emergency managers can better serve those with critical needs and create communities that are more resilient.

Mary Lasky, a Certified Business Continuity Professional (CBCP), serves as the program manager for business continuity planning for the Johns Hopkins University Applied Physics Laboratory (JHU/APL), where she coordinated the APL Incident Command System Team. She also as a member of InfraGard, where she is on the executive committee for the InfraGard EMP-SIG; and the Federal Emergency Management Agency's Nuclear-Radiation Communications Working Group. In Howard County, Maryland, she serves as president of the Community Emergency Response Network Inc. (CERN); president of the board of directors of Grassroots Crisis Intervention Center; and Finance Committee member for Leadership Howard County and is co-chair of the Steering Committee for the Leadership Premier Program. For many years, she has been on the adjunct faculty of the Johns Hopkins University Whiting School of Engineering. She is the immediate past president of the Central Maryland Chapter of the Association of Contingency Planners (ACP) and has held a variety of supervisory positions in information technology and in business services. Her consulting work has included helping nonprofit organizations create and implement their business continuity plans.

Mass Group Resilient Hospital in Rwanda

The following article from the Mass Group website provides background to the presentation made by Sierra Bainbridge on the resilient hospitals panel.

https://www.massdesigngroup.org/portfolio/butarohospital/

"The new facility, which opened in January 2011, is designed to mitigate and reduce the transmission of airborne disease through various innovative systems, including overall layout, patient and staff flow, and natural cross-ventilation.

To design the hospital we made use of local materials—such as nearby volcanic rock from the Virunga Mountains—and labor intensive practices to deliver appropriate and sustainable design as well as stimulate the local economy through employment. This approach reduced the facility's price tag to roughly two thirds of what a comparable hospital would typically cost in Rwanda, saving 2 million dollars in construction fees but also providing over 4,000 jobs. The Butaro Hospital brought together architects, builders, and doctors directly to a community in need, providing a dignified, healing space for its population and solidifying for MASS the value of purposeful, human-centric design.

This project was completed in collaboration with Partners In Health and the Rwanda Ministry of Health."

Partial reproduction of the New York Times article here:

IN 2006, a 26-year-old architecture student, "Michael Murphy, approached the global health pioneer Paul Farmer after a lecture at Harvard. Mr. Murphy asked which architects Dr. Farmer had worked with to build the clinics, housing, schools, and even the roads he had described in his talk. An aspiring social entrepreneur, Mr. Murphy was hoping to put his design degree to use by apprenticing with the humanitarian architects aiding Dr. Farmer's work. But it turns out, those architects didn't exist.

"I drew the last clinic on a napkin," Dr. Farmer told Mr. Murphy.

Soon after, Mr. Murphy flew to Rwanda, where he and a few other students, including Alan Ricks and Marika Shioiri-Clark, became Dr. Farmer's architects. Mr. Murphy lived in the country for over a year while the Butaro Hospital, which laborers built with local materials, was designed. Now, a site that was once a military outpost is home to a 150-bed, 60,000-square-foot healthcare center that served 21,000 people in its first year and currently employs 270, most of them locals in an area with chronic unemployment.

The Butaro Hospital is a breathtaking building with intricate lava rock walls made of stones cut by Rwandan masons, and it is full of brightly colored accent walls and breezeways bathed in light and air. Deep-green flora blossom everywhere. For the 340,000 people who live in this region of Northern Rwanda, the project marks a literal reclamation: an area that was once a site of genocidal violence is now a center for state-of-the-art medical care. Healing happens there. An unmistakable grace permeates the place.

Building the hospital under the auspices of the nonprofit MASS Design Group (MASS stands for a Model of Architecture Serving Society), Mr. Murphy, Mr. Ricks, and Ms. Shioiri-Clark relied on Dr. Farmer's theory of a "preferential option for the poor." The idea—adopted from liberation theology—is that the poor deserve the best quality intervention because they've been given the least by luck and circumstance. The students' naïve audacity, coupled with Dr. Farmer's wisdom and experience, resulted in a building that has set a new standard for public-interest design.

It used to be that young people with humanitarian aspirations went into law or medical school or applied to Teach for America or the Peace Corps. But today, increasing numbers of the most innovative change makers have, like Mr. Murphy, Mr. Ricks, and Ms. Shioiri-Clark, decided to try to design their way to a more beautiful, just world.

Though unemployment is widespread among designers and architects, there exists a world of products, places, and processes in desperate need of redesign. Imagine if designers—uniquely trained to listen and observe, and to improve the way things function, feel, and look—were, like the Enterprise Rose fellows, embedded in schools, community centers, nonprofit organizations, health clinics, religious institutions, and government offices, where they could experience community needs and behavioral patterns firsthand.

The need for designers—and their ingenuity and interest in beauty and functionality— is not limited to Africa, India, Haiti, or other far-flung places where architects and designers are commonly called upon following natural disasters. People who struggle to maneuver strollers and wheelchairs in and out of urban transportation systems or work in a deadening sea of suburban office complexes share the same basic need for enlivening, dignifying design. Anyone who has recently visited a local motor-vehicles office most likely knows about the need we have in mind.

Mr. Murphy had a surprising insight about how much the developed world has to learn about good, human-centered design from the developing world. After finishing the Butaro Hospital and returning to the United States, Mr. Murphy said, he was struck "at how over-designed most hospitals are here—yet there's little natural airflow, a lack of color and craftsmanship, and few outdoor spaces to take a deep breath and gain some perspective."

When faced with a poorly considered, dehumanizing product—be it a dingy women's center, a mountain of unnecessary bureaucracy, or assembly instructions for a new product that make you feel inept—it is a failure of design. The bad news is that no country, rich or poor, is immune to bad design; the good news is that we can all learn from one another.

But we have to advocate for it and many of us, until now, simply haven't realized that we deserve better. We couldn't imagine the alternative. But once you see what good design can do, once you experience it, you can't unsee it or inexperience it. It becomes a part of your possible. The public-interest design movement is counting on it."

Testimony of James W. Terbush MD, MPH
To the Blue Ribbon Study Panel on Bio-Defense

The Hudson Institute, Washington DC
April 1, 2015

Thank you, Mr. Chairman for asking me to take part in this important panel. Up until April of last year, I was actively involved in domestic disaster preparedness and response for the military working at NORAD and U.S. Northern Command. Prior to that, I was fortunate in my career to be able to participate in International Disaster Preparedness and Response, specifically the 2010 Haiti Earthquake response. Since leaving Government, I have enjoyed teaching disaster public health and consulting on a variety of health- and disaster-related topics. I am delighted to be able to speak with you today on a topic I consider to be of key importance to our nation, especially when we experience another bio-event such as Ebola or other significant natural or man-made disaster, that is, Resilient Hospitals.

Resilient Hospitals: In the opening chapters of the celebrated book, Five Days at Memorial, the author Sheri Fink recounts in detail the horrifying facts of "life and death in a storm-ravaged hospital" post-Hurricane Katrina. She describes a major medical center without electricity, clean water, waste-water treatment, ventilation, and only limited communications, supplies, and transportation. Patients, deprived of lifesaving technology, lingered and then died in the heat, a nightmarish scenario indeed. Of the 16 critical infrastructure sectors, Healthcare and Public Health is certainly important in immediate disaster response and recovery. The population we serve, those critically ill and injured hospitalized patients, are arguably the most vulnerable segment of our society. The other reason perhaps we need to focus on resilient hospitals today is that the sector is one of increasing complexity and relies on a combination of support from the other sectors, especially the power grid and increasingly a reliance on moment-to-moment connectivity with Information Technology (IT) and the Internet.

Resiliency: For our purposes, a working definition of resiliency is "the ability to take a blow and come back". Resilient Hospitals are able to prepare for and adapt to changing conditions and withstand and recover rapidly from disruptions. Not just hospitals should be considered but healthcare facilities of all types are vulnerable and need to be "resilient". Patients receive care at a variety of different facilities to include long-term care (nursing homes) and clinics. It is not just the physical structure which must withstand a blow and come back, but we need resilient staff, resilient management, resilient plans, and planning. A Naval Aviator friend told me once that "Truly superior pilots plan ahead to avoid those situations where they might have to use their superior skills". Hospitals and their staffs have repeatedly shown superior skills in disasters, but we would prefer to have less heroics and more routine activity carried out "according to the plan".

Hospitals have unique vulnerabilities: Patients are more sensitive to changes in environment; temperature, humidity, noise, etc. The very young and the very old (and of course the very sick) often have very different requirements. Some patients have to be isolated from others and need separate ventilation systems, changing rooms, etc. Some patients rely on ventilators or other specialized technology with a limited battery supply of perhaps several hours when the power goes off. If back-up diesel generators are tasked to perform beyond their limits, these devices eventually run down as well. Another vulnerability common to devices connected to the Internet is that some medical devices, to include some life-sustaining medical devices can be "hacked" remotely, either turned off or the settings changed. When the power goes completely off, hospitals quickly become dark and dangerous places. Most back-up electrical generators are designed for no more that 48–72 hours of continuous operations. They probably need routine maintenance and certainly re-fueling. Another unique issue is the evacuation of critically ill patients connected to life-support to another healthcare facility. When only one or two facilities are no longer able to care for patients, the option remains to evacuate. When healthcare facilities across an entire region are affected, we have to be able to continue to provide for these patients in place.

Electronic medical records and data stored in the "Cloud" can be both a help and a hindrance in a disaster: When IT systems and access to the Internet stop, modern medicine, as we know it, ceases to exist. Although valuable patient records and other data may exist somewhere "out there" on a server, inability to access or retrieve data stops our business as usual. The ability to record and store patient demographic and clinical information on a secure hand-held device, especially in a mass casualty, is essential. The data can be downloaded later or sent to another device when Internet connectivity is restored. Tracking of patients and accompanying family members is particularly important when facilities are being evacuated. Otherwise, we may have to revert back to paper records and clipboards, which were both used in the mass shooting incident in Aurora Colorado, a couple of years ago. Less effective perhaps, but that ability to revert back to another legacy system is also an indicator of resiliency.

A "just in time" supply chain complicates disaster healthcare delivery: The need for cost-effectiveness complicates resiliency. Because it is more cost-effective to have vendors deliver supplies "just in time" there is less waste and less wasted shelf space. The days of large stocks of IV fluids, pharmaceuticals, and "disposables" are gone. Instead vendors may obtain supplies from multiple sources, both domestic and overseas, and those vendors in turn have a supply chain from even more obscure sources. IT systems connect them all. As systems become increasingly complex, they are often increasingly fragile. For no-fail missions (such as disaster healthcare, communications, ICU/CCU's, life-support, and emergency rooms), we need redundancy and additional capability. This may include more trained staff, equipment, and supplies in-house. Functions within the hospital are then prioritized as mission critical, or non-mission critical such as we do in the military. Functions of lower priority may then need to be turned-off in an orderly manner, as in the phrase "failing gracefully". All this adds to business costs; staff hours, overhead, liability, and represents an additional "risk" to the hospital.

Alternate technologies can be useful in disaster, if they are "baked in": PPD -21 promotes research and development to enable the secure and resilient design and construction of critical infrastructure and more secure accompanying cyber-technology. An architectural firm based in Boston is designing hospitals "from the ground up" which have more natural ventilation and lighting, are more sparing in the use of water and have a reduced requirement for waste-water treatment. Some of these hospitals include a thermal tower which pulls air through the facility without electricity. They have large fans in common areas in case this does not work. The day-to-day electricity requirements for these hospitals are much less, more of the hospital is on the ground level and patients can be moved more easily without the use of elevators. Why is this technology not used more commonly? Because these hospitals are being designed for third world situations in Africa and elsewhere. Certainly, these countries who experience disasters and loss of life more frequently than we do here, benefit from this technology. Maybe certain adaptations of this type of technology are needed here to make our hospitals more resilient? These technologies are appropriate and resource-saving all of the time and do not have to be "turned on" in disaster.

Cyber-secure micro-grids: Another example of technology useful "all of the time" is the use of a back-up electrical generation system, incorporating conventional diesel generators, renewables, batteries, and the ability to push power back into the grid, with possible associated cost savings. These micro-grids are less susceptible to hack attacks and electromagnetic disturbances (EMP) as well. I've seen one of these systems seamlessly transition from providing power for a large portion of a military base, to putting power back into the grid, and then storing energy in batteries. They also have the ability to have parts of the system go offline for maintenance or re-fueling, which was a problem we saw in Super-storm Sandy. Currently, the Department of Defense has such a Joint Capability Technology Demonstration (JCTD/SPIDERS) which could be adapted for use in a large medical campus, for example. I can discuss this subject of cyber-secure micro-grids in greater detail if you are interested.

A "System of Systems" approach is needed: A favorite slide I like to use when giving similar talks is one showing the critical infrastructure sectors, stacked on top of each other, with lines of interconnectedness. The power grid relies upon transportation; transportation is connected to water; the water sector needs electricity; and IT seems embedded in all sectors. Healthcare may not directly affect all the other sectors but it is fair to say all the other sectors affect healthcare. Especially vulnerable in a disaster are patients at home or in a long-term care facility which must have an electrical outlet for life-sustaining technology; ventilators, oxygen generators, and renal dialysis.

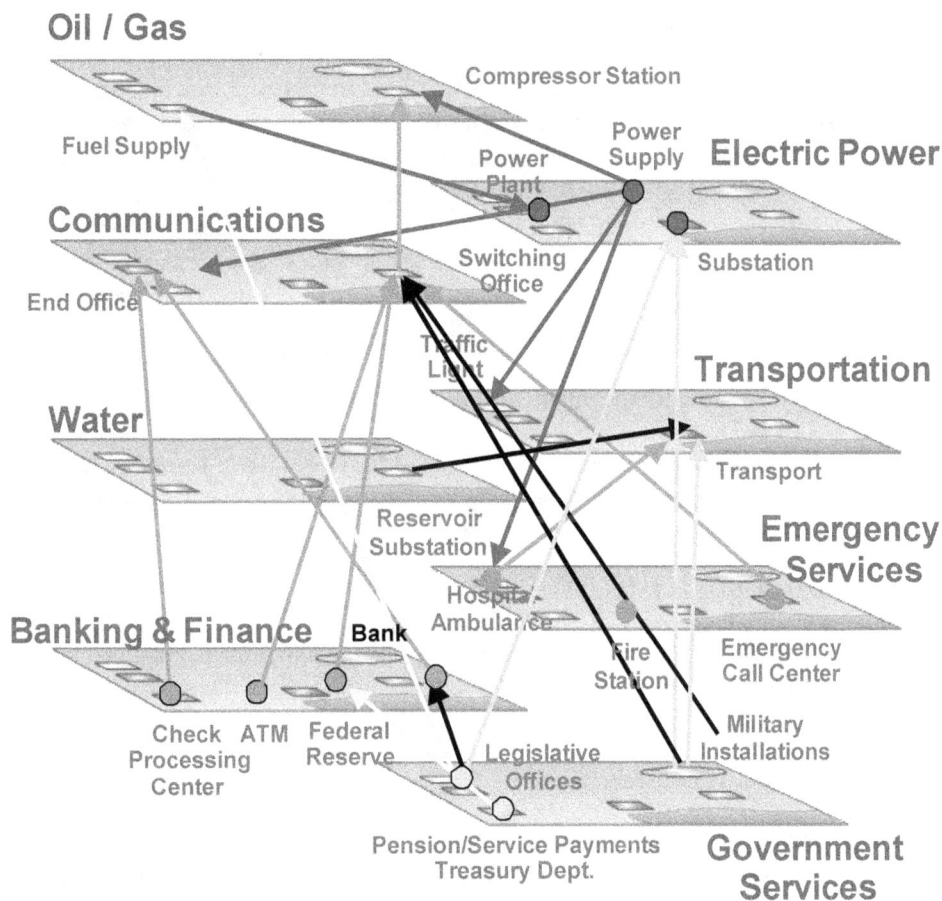

Changes to policy have benefit: One way to promote some of these changes is through legislation, rulemaking, and using the recertification process for hospitals. Hospital preparedness funding from Government has been limited when compared to the magnitude of the healthcare industry (estimated now at $2.9 Trillion annually). But regulation (and some say overregulation) represents an additional "risk" to hospitals in terms of costs. Preparedness needs to make business sense too; continuity of operations, hospital reputation in the community, protection of staff and their families as well as patients. Hospitals are frequently looked at as a community resource for the moderately ill and elderly, a place of safety and help, and a place to go to in disaster. This complicates the decision-making process when there are already limited resources available. Coalitions of Hospitals, as we heard from a previous speaker (Tim Stephens from MESH), make sense when it comes to business continuity and preparedness. Grant funding from the Government will never be enough by itself (a colleague told me you can never "Grant" your way to preparedness). Specific funding directed at a specific type of vulnerability, such as hospital cyber-security or emergency power supplies, can however help.

How we educate healthcare professionals: One of the most satisfying professional tasks I have is to teach a course on emergency public health. Disaster public health or "Disaster Medicine" is an essential part of a healthcare provider's training. Any healthcare professional can be faced with providing services in an emergency. No specialty is completely exempt from responsibility to learn about disaster health, even if it is only to know your own place in the response system. Disaster Medicine is closely related to the larger field of study, which is Public Health. The concepts of population health; vulnerable populations, determinants of health, and at-risk groups are fundamental. A course in disaster healthcare needs to be in all medical and nursing schools' curriculum, and more focused training needs to be available to providers already in the community as a part of continuing medical education and/or professional licensure.

A culture of preparedness: Finally, individuals and families need to take on more responsibility for their own needs in advance of disaster. This includes a family disaster plan. For obvious reasons, the fewer the persons needing assistance from hospitals for disaster-related illness and injury, more of the limited medical resources can be directed toward those in greatest need. Despite numerous examples and warnings, individuals and families still fail to prepare for even "predictable surprises", such as power outages, stores closing, and bad weather. The normalcy bias or "fallacy of normalcy" as one author put it, makes us assume that tomorrow will be very much like today and what we are seeing on radio, television, or hearing from neighbors is exaggerated and causes some of us to drastically underestimate the effects of the disaster.

Thank you for the opportunity to address this prestigious panel on Resilient Hospitals.

Note. The U.S. Department of Health and Human Services has awarded more than $840 million to continue improving emergency preparedness of state and local public health and healthcare systems. These systems are vital to protecting health and saving lives during a disaster.

The grant funds are distributed through two federal preparedness programs—the Hospital Preparedness Program (HPP) and the Public Health Emergency Preparedness (PHEP) programs. These programs represent critical sources of funding and support for the nation's healthcare and public health systems. The programs provide resources needed to ensure that local communities can respond effectively to infectious disease outbreaks, natural disasters, or chemical, biological, or radiological nuclear events.

The fiscal year 2014 funding awards include a total of $228.5 million for HPP and $611.75 million for PHEP.

HPP funding supports building sustainable community healthcare coalitions that collaborate on emergency planning and, during disasters, share resources and partner to meet the health and medical needs of their community.

PHEP funding is used to advance public health preparedness and response capabilities at the state and local level.

HPP and PHEP funding helps recipients build and sustain public health and healthcare.

References:

Beyond Hurricane Heroics, Sheri Fink. http://sm.stanford.edu/archive/stanmed/2013summer/article5.html.

Five Days at Memorial. http://www.randomhouse.com/book/203207/five-days-at-memorial-by-sheri-fink.

PPD-21. https://www.whitehouse.gov/the-press-office/2013/02/12/presidential-policy-directive-critical-infrastructure-security-and-resilience.

http:/www.dhs.gov/critical-infrastructure-sectors.

http:/www.dhs.gov/healthcare-and-public-health-sector.

Healthcare Costs in the United States. 2013. http://www.cms.gov/Research-Statistics-Data-and-Systems/Statistics-Trends-and-Reports/NationalHealthExpendData/NationalHealthAccountsHistorical.html.

Cyber-secure Micro-grid. http://energy.gov/eere/femp/articles/spiders-joint-capability-technology-demonstration-industry-day.

See also: Barnett, Daniel J., Tara K. Sell, Robert K. Lord, Curtis J. Jenkins, James W. Terbush, and Thomas K. Burke. "Cyber Security Threats to Public Health." World Medical & Health Policy 5 (1) (2013): 37-46.

The original publication can be found here: http://www.domesticpreparedness.com/Commentary/Viewpoint/ Talking_to_People_Who_Do_Not_Believe_Bad_Things_Can_Happen/

Talking to People Who Do Not Believe Bad Things Can Happen

by WILLIAM KAEWERT

Wed, February 25, 2015

Soldiers, law enforcement officers, emergency responders, and others whose professions involve responding to or mitigating catastrophic events tend to think about "bad things" more often than the average person because they either deal with life-and-death issues regularly or have received training to do so. The term "bad things" in this article refers to high-impact threats to the well-being of a large number of people in a wide area—for example, any natural disaster or deliberate attack with the potential to cause cascading infrastructure failure. When receiving bad news, there are different ways in which less concerned people can unrealistically minimize threats, which include but are not limited to:

- Those who believe that bad news happens all the time and, as a result, they may tune out the media.

- Those inexperienced with disasters and, therefore, do not believe until it is too late that they could be affected.

- Those who believe that, if the situation deteriorates, the government will take care of them.

Common Barriers to Communication: Each of the above behaviors or attitudes can cause the people who embrace them to be unprepared for disasters. Communicating about high-impact threats to people who do not want to hear about them can be a challenge. However, there are common-sense approaches that can improve the odds of successful communication and perhaps lead to positive action. The benefits of communicating about bad things accrue to both parties. Becoming more self-sufficient enables citizens to better endure disasters while experiencing less stress. Emergency responders benefit because a well-prepared citizenry means reduced demand for emergency services during a disaster.

When confronted with a new problem or threat for the first time, some people may become defensive. At an EMPact America conference in September 2009, Peter Huessey, senior defense consultant of the National Defense University Foundation, described four types of barriers people erect when confronted with new information: (a) "Not invented here" (distrustful attitude); (b) "How often has that happened?" (sarcastic attitude); (c) "What are you selling?" (skeptical attitude); and (d) "How come I haven't heard of this before?" (defensive attitude).

These barriers may be conveyed by words or body language and include an underlying attitude behind the behavior. There are effective techniques for addressing these types of resistance. Depending on the situation (group presentation, tabletop exercise, or one-on-one discussion), one or more of the following approaches may be helpful in breaking through the other party's preconceived notions that underlie their defensiveness.

Tell a Story: Personal stories about problems and how they affect presenters and listeners often are more effective than lectures for communicating a concept about which the person may not be an expert, such as a politician discussing the possibility of an extended power outage. Conveying that a friend's wife would die without her diabetes medication more powerfully illustrates the problem of an extended power failure than lecturing about maintaining sufficient reserve supplies. Personal stories combined with sincere feelings help listeners relate to the presenter, thus reducing the chance of confrontation.

Be Credible: Allaying fears about the presenter's motives can improve the relationship between presenter and participants by reducing suspicion of a hidden agenda. If the target audience does not already know the presenter well, providing them with information about the presenter's background, training, and organizational affiliation boosts credibility, as does telling the truth, preparing thoroughly, and attributing all research material to relevant sources.

In addition, audiences often relate well to presenters who explain from whom they learned about a particular problem, display an appropriate level of humility, such as admitting when they do not have answers, and refraining from telling people that everything "is under control" or "will be all right" when no such assurance is possible. Sometimes listeners feel embarrassed when they think they know less than others and, as a result, may act defensively. Presenters can help listeners overcome this hurdle by explaining they once did not know about the threat being discussed, and sharing where they learned of it.

Choose Wisely: An obvious example of the wrong time to initiate the subject of catastrophic threats is at a cocktail party, where people reasonably expect to relax and unwind. The chance of having a successful conversation about bad things increases when saying the right thing to the right people at the right time. Three points to consider are the following:

- *Timing*—Initiate conversations when the audience sends clear signals that it is receptive, not when the presenter feels like talking.

- *Research*—Understanding the audience can pay big dividends by helping a presenter tailor an appropriate message. Presenting a disaster scenario that fits the listener's worldview, for example, can reduce the problem of listeners "tuning out" the presenter.

- *Discernment*—Sometimes presenters face unexpectedly difficult listeners. A shrewd presenter asks questions to discern the listener's motives, and adjust his or her approach accordingly—including disengaging from people that the presenter's information will not help.

Get Help: There is a wealth of published information about the causes of, preparation for, and recovery from nearly any disaster imaginable. Reports are available from government, nonprofit, university, think-tank, corporate, and other sources. Some carefully researched novels based on a variety of disasters from financial system meltdown to electromagnetic pulse attack can be powerful triggers of the imagination. Leaving trustworthy reading material behind after an exercise or presentation can reinforce the message.

Ask for Action: After successful communication, the next step is to ask for action, such as developing an emergency plan, writing to elected representatives, or improving neighborhood relationships. The earlier this goal is defined when planning any interaction or presentation with the target audience, the more likely the goal will be achieved. The primary goal of such presentations is to help people imagine what a disaster would mean for them and encourage them to respond by taking small steps toward becoming more self-sufficient. As their preparedness grows, they will be in better shape when disaster strikes and less of a burden on emergency-response systems that could well be overstressed during the next "bad thing."

William Kaewert is the founder of two power protection companies and has over 30 years of experience applying technology-based solutions that assure continuity of electrical power to critical applications. He is currently president and chief technology officer of Colorado-based Stored Energy Systems LLC (SENS), an industry-leading supplier of nonstop DC power systems essential to electric power generation and other critical infrastructures. The company also produces commercial off-the-shelf (COTS)-based power converters used in military systems hardened for electromagnetic pulse (EMP), including ground power for the Minuteman III ballistic missile system and the Terminal High Altitude Area Defense (THAAD) missile interceptor. He received his AB in history from Dartmouth College and MBA from Boston University. He serves on the board of directors of the Electrical Generation Systems Association and on the management team of the FBI InfraGard Electromagnetic Pulse Special Interest Group.

Why Bother to Prepare?—A Government Website Illustration of the Unintended Consequence of Neglecting High-Impact Disaster Scenarios

By Charles Manto

The following ready.gov website was live as of September 2015. It shows the importance of creating a checklist and having food that is not dependent on refrigeration because of a power outage that can accompany a disaster. However, only common short-term duration disasters are considered and there is no mention of high-impact low-frequency disasters. "Power outages … could last for several days." This exclusive focus on short-term disasters unintentionally undermines preparedness since many could assume that they could survive without a few days of food and expect to be rescued by day 4. Especially when times are hard, some might ask, "Why take time and money to be prepared when we will be rescued in a few days anyway?" However, placing preparedness and disaster planning in a complete context of disasters including long-term regional and national disasters that include long-term power outages would encourage greater resilience at the personal and community level. It would also foster a greater sense of engagement and responsibility on behalf of the entire community and lessen an overdependence on being rescued from some group outside the disaster area, which may or may not materialize quickly in the event of a nationwide disaster.

http://www.ready.gov/food

FOOD

Recommended Supplies List

Consider the following things when putting together your emergency food supplies:

- Store at least a three-day supply of non-perishable food.

- Choose foods your family will eat.

- Remember any special dietary needs.

- Avoid foods that will make you thirsty.

- Choose salt-free crackers, whole grain cereals, and canned foods with high liquid content.

Following a disaster, there may be power outages that could last for several days. Stock canned foods, dry mixes, and other staples that do not require refrigeration, cooking, water, or special preparation. Be sure to include a manual can opener and eating utensils.

BASIC DISASTER SUPPLIES KIT

Recommended Supplies List

A basic emergency supply kit could include the following recommended items:

- Water, one gallon of water per person per day for at least three days, for drinking and sanitation.

- Food, at least a three-day supply of non-perishable food.

- Battery-powered or hand crank radio and a NOAA Weather Radio with tone alert and extra batteries for both.

- Flashlight and extra batteries.

- First aid kit.

- Whistle to signal for help.

- Dust mask to help filter contaminated air and plastic sheeting and duct tape to shelter-in-place.

- Moist towelettes, garbage bags, and plastic ties for personal sanitation.

- Wrench or pliers to turn off utilities.

- Manual can opener for food.

- Local maps.

- Cell phone with chargers, inverter, or solar charger.

- Additional Emergency Supplies.

- First Aid Kit.

- Supplies for Unique Needs.

Once you have gathered the supplies for a basic emergency kit, you may want to consider adding the following items:

- Prescription medications and glasses.

- Infant formula and diapers.

- Pet food and extra water for your pet.

- Cash or traveler's checks and change.

- Important family documents such as copies of insurance policies, identification, and bank account records in a waterproof, portable container. You can use the Emergency Financial First Aid Kit—EFFAK (PDF—977 Kb) developed by Operation Hope, FEMA, and Citizen Corps to help you organize your information.

- Emergency reference material such as a first aid book or free information from this website (see Publications).

- Sleeping bag or warm blanket for each person. Consider additional bedding if you live in a cold-weather climate.

- Complete change of clothing including a long sleeved shirt, long pants, and sturdy shoes. Consider additional clothing if you live in a cold-weather climate.

- Household chlorine bleach and medicine dropper—When diluted, nine parts water to one part bleach, bleach can be used as a disinfectant. Or in an emergency, you can use it to treat water by using 16 drops of regular household liquid bleach per gallon of water. Do not use scented, color safe, or bleaches with added cleaners.

- Fire extinguisher.

- Matches in a waterproof container.

- Feminine supplies and personal hygiene items.

- Mess kits, paper cups, plates, paper towels, and plastic utensils.

- Paper and pencil.

- Books, games, puzzles, or other activities for children

This essay published in the November 2014 issue of DomesticPreparedness.com and the full DomPrep issue named Blackout addresses issues discussed by the several 2014 Dupont Summit panels including two the author joined covering legal and regulatory issues. The author steps through her experiences proposing successful legislation in Maine. Updated status of the legislation and preparedness steps in Maine will be discussed in the 2015 Dupont Summit. This article is reprinted with permission from the IMR Group, Inc., publisher of DomesticPreparedness. com, the DPJ Weekly Brief, and the DomPrep Journal. The IMR, Inc. offers no guarantees as to the accuracy of any information presented, but encourages all readers to use IMR, Inc. programs primarily as a resource to facilitate their own research.

The original article can be found here: http://www.domesticpreparedness.com/Government/State_Homeland_ News/Maine_-_A_Journey_Through_State_Grid-Protective_Legislation_%26_the_Threat_of_Regulatory_ Capture/

Maine—A Journey Through State Grid-Protective Legislation and the Threat of Regulatory Capture

by ANDREA BOLAND

Wednesday, November 19, 2014

What happened in Maine when the state legislature, receiving testimony from national experts, resolved to protect the electric transmission system from severe geomagnetic disturbances (GMD) and manmade electromagnetic pulse (EMP) weapons is a study in the stresses imposed by appointed regulatory bodies on legislative policymaking bodies. Here is a sketch of actions by the state and electric utilities, operating at both the federal and state levels.

The Maine Legislation: Legislative Document 131 (LD 131), "An Act to Secure the Safety of Electrical Power Transmission Lines," initially required all current and future transmission system upgrades to include protections against both solar storms (GMD/geomagnetic disturbance) and manmade electromagnetic pulse (EMP) weapons and terrorist devices. The Joint Standing Committee on Energy, Utilities, and Technology (EUT) of the Maine State Legislature held the public hearing in February 2013, and work sessions thereafter.

The electric industry initially opposed the legislation claiming that it was not needed. After compelling, data-driven testimony by independent experts showing big gaps in the security of the Maine grid, the EUT decided the whole grid needed to be protected. They turned LD 131 into a "resolve" that required the Maine Public

Utilities Commission (PUC) to examine the vulnerabilities of the transmission system and identify options for mitigation measures—including low-, mid-, and high-cost options, and a time frame for adoption.

The committee approved the bill unanimously as "emergency" legislation. Then, the House approved LD 131 unanimously; the Senate approved by a 32-3 vote; and the resolve became law on June 11, 2013. Its preamble states:

> **"Whereas**, in the judgment of the Legislature, these facts create an emergency within the meaning of the Constitution of Maine and require the following legislation as immediately necessary for the preservation of the public peace, health and safety; now, therefore, be it." (See the full text of the June 2013 Maine legislation on GMD and EMP)

It was a very clear directive. The Maine legislation called for a report from the Maine PUC due by January 20, 2014. Thomas Welch, chair of the Maine PUC, anticipated an on-time report; he noted that, as chair of the PUC, he could approve the reliability upgrades without awaiting the report. Representative Barry J. Hobbins (D-Saco) had promised at the close of the public hearing of the EUT, "I don't know what we are going to do, but I can tell you this—we are going to do something!" The legislative intent was clear: provide the information needed to protect Maine's electric grid. This was landmark legislation, heralded nationally and internationally. A single state had done what Washington DC, never has—passed legislation for GMD and EMP protections.

Inadequacy of Federal Protection: Efforts seeking GMD and EMP grid protections at the federal level have been frustrated. The Federal Energy Regulatory Commission (FERC) has no legal authority to initiate "reliability standards" for the electric utilities. Only the North American Electric Reliability Corporation (NERC), the industry association, has that authority.

When FERC tells NERC they must set those standards, NERC writes weak standards, say the standards are not needed, or argue for more time. With effective lobbying, the utility industry has repeatedly blocked federal legislation that would give FERC power to require higher reliability standards than those NERC proposes.

The electric utilities comprise the only national infrastructure that is self-regulated. Some independent experts worry about the degree to which FERC seems to accommodate them. William R. Harris, secretary and counsel to the Foundation for Resilient Societies, on November 1, 2014, compared the regulatory capture problem at FERC with that at the Federal Reserve in a shared email among interested parties, responding to a publication article.

> *What a contrast: the Fed [Federal Reserve Bank] has extraordinary information subpoena powers, sanctions authority, and standard-setting that is not subject to veto by the regulated banks. Yet the Fed acts as if they must "get along" with the regulated firms even when they place society at great risk.*

> *A fortiori, far weaker FERC Commissioners act as if they need to ingratiate themselves to the electric utility industry that operates with monopoly power to initiate reliability standard-setting.*

> *The behavioral aspects of "regulatory capture" appear paramount. With FERC, Cheryl LaFleur (Chair of FERC) acts as if she has a psychological "need" to be in sync with the NERC culture, and to act as if scientifically-defective NERC reliability standards are OK because, as members of the NERC Board keep repeating: "Reliability is in our DNA."*

The key difference between FERC and the Maine PUC is that the PUC, like the Federal Reserve Bank, has the authority to require the utilities to employ specific protections, even without waiting for a study. Also, like the Federal Reserve Bank, the PUC has come under scrutiny by Maine's Government Oversight Committee for a possible "culture problem," otherwise described above as "regulatory capture."

Protecting Maine's transformers, which would be irreplaceable for years in the event of a widespread blackout, conservatively is estimated at $7.2 million, or about $2.80 per household per year for about five years. If shared across all the states within the target area of ISO New England—the independent, not-for-profit company authorized by FERC to perform grid operation, market administration, and power-system planning roles for the region—that cost would drop to about $0.35. The protective equipment should last at least two decades, so the average cost over the life of the equipment would then be either $0.70 or $0.09 per household, respectively.

The Maine Public Utilities Commission Takes Over: From a hopeful start, in which the PUC provided an online docket for the study (Maine PUC docket 2013-00415, available online), it devolved into a draft report in December 2013 that showed utility bias, recommending: do nothing; wait for Washington. Reaction to the report was quick, professional, and condemnatory. National experts again turned their attention to the report and detailed errors, omissions, and shortcomings, and made corrective recommendations. Dr. Peter Pry, executive director of the Task Force on National and Homeland Security, conducted analysis that found the report to be "dishonest." His analysis is on the online Maine PUC docket 2013-00415.

When Welch presented the final report to the EUT in January 2014, he acknowledged GMD and EMP were serious problems, hence the PUC needed more time. Delay was reminiscent of NERC/FERC history. Welch proposed a task force with Central Maine Power Company (CMP) as coordinator. CMP had a convenient location, but had been unable to answer legislator questions during hearings. At one point, Chair Hobbins said, "I feel if we ask you one more question, you'll throw up your hands and say, 'guilty.'" The sponsor and experts welcomed the revived focus, and inclusion of outside expertise.

Two independent experts were invited to participate: John Kappenman of Storm Analysis Consultants; and Thomas Popik, president of the Foundation for Resilient Societies. EMPrimus, a research and development company that had offered expert docket input, also was invited to participate. The monthly meetings focused predominantly on GMD. CMP's Brian Huntley's leadership seemed strong, but there was some worry that, as with NERC/FERC processes, the report would be sabotaged before it was complete.

In September 2014, Huntley left CMP, and the task force meeting was canceled. On September 24, Welch assured the Government Oversight Committee that the report would be out by December, EMP would be covered, and staffers were working with EMPrimus. Later that day, he announced his early retirement for December 31, 2014. [Note: As the bill's sponsor, I made phone calls to CMP that were not answered.]

Central Maine Power Company Finds No Protective Equipment Is Needed—The Grid Could Withstand Any Threat Their Model Could Conceive: The last study group meeting was on October 27, 2014. Justin Michlig, lead engineer of system planning at CMP, who became the new CMP project manager of the study in late September, invited EMPrimus to present its report. EMPrimus utilized independent (PowerWorld Corporation) modeling, real-world historical data, mitigation equipment, and NASA probabilities for a 100-year solar storm (12 percent within a decade, 50 percent within a 30-year period). The EMPrimus report found that "reactive power" equipment might temporarily stabilize grid voltage in a solar storm. However, there would be five-minute periods necessary to reset key equipment, during which the grid would be at risk of collapse. During prior solar storms, Maine's reactive power equipment had become inoperable on multiple occasions. EMPrimus proposed the installation of 18 neutral ground blocking devices. These would protect transformers and keep geomagnetically induced currents out of the high-voltage Maine transmission system.

Then, Michlig presented CMP's analysis. The modeling relied on the technically dubious NERC "GMD Benchmark Event" methodology in the still unapproved standard and did not use CMP's own recorded data for validation. The CMP scenario assumed that "reactive power" equipment always worked, despite outages in past solar storms. CMP found no need to install any protective equipment. Michlig gave no answers to questions of why they ignored their own historical data. Lisa Fink, PUC staff attorney and project manager of the PUC work, backed up Michlig. The CMP Draft Report will be available later in November 2014 for comment.

It was surreal, appearing that Maine's PUC exhibited the kind of "regulatory capture" that has troubled the Maine public and its legislature. Like NERC, the PUC had cordially supported the mission, opened an online docket, and then manipulated data and assumptions to avert protection of the Maine grid.

The emergency legislation did not ask for the recommendation of the Maine task force and its PUC staffers. It asked them to identify vulnerabilities, options for protections, and costs, so that the legislature, on behalf of the people of the State of Maine, could make their own decisions on protections. [Note: As sponsor, I reminded them of that at the meeting and asked them to be sure to read the legislation; they said they would.]

At the time of this writing, the nation still awaits the draft report. However, the PUC has taken down the online docket—a recent change that blocks public view.

_____ Update (11/21/14): Representative Boland, working with others, was able to get the docket back up online.

State Representative Andrea Boland is completing her eighth year in the Maine legislature. She is considered a leader in safety issues of electromagnetic radiation, especially from cellphones and smart meters. She became involved in electric grid protection against electromagnetic pulse and geomagnetic solar storms (GMD) at the suggestion of her regular scientific advisor. Her work is supported by several national experts. She earned a BA degree from Elmira College and an MBA from Northeastern University, and studied at the Sorbonne and Institute of Political Studies in Paris. She was awarded the 2011 Health Freedom Hero Award by the National Health Federation for her work on health freedom and safety. Her legislative work has led to confronting major corporate interests on matters of transparency and regulatory capture, and public protections.

IS YOUR CRITICAL NETWORK INFRASTRUCTURE AT RISK OF A SECURITY BREACH?

YES!

WHY?

Hackers know how the network security game is played...

CHANGE THE GAME!

Avaya Fabric Connect offers solutions delivering network virtualization, cloud, mobility, and video—while simplifying the network, increasing deployment speed and building in higher levels of security.

AVAYA
Engage The Power of We™

avaya.com/fabricconnect

This essay published in the Viewpoint section of DomesticPreparedness.com addresses issues discussed by the several 2014 Dupont Summit panels including two the author joined covering legal and regulatory issues. It explains how complex and lengthy the legislative process can become and illustrates why neutral forums such as InfraGard's EMP SIG and the Dupont Summit are very helpful in bringing various viewpoints on tech policies to light. This article is reprinted with permission from the IMR Group, Inc., publisher of DomesticPreparedness.com, the DPJ Weekly Brief, and the DomPrep Journal. The IMR, Inc. offers no guarantees as to the accuracy of any information presented, but encourages all readers to use IMR, Inc. programs primarily as a resource to facilitate their own research.

The original publication can be found here: http://www.domesticpreparedness.com/Commentary/Viewpoint/ Political_Realities_of_Legislation_for_Extreme_Events/

Political Realities of Legislation for Extreme Events

by ANDREA BOLAND

Wednesday, September 16, 2015

The single extreme solar storm (GMD/geomagnetic disturbance) or electromagnetic pulse (EMP) attack (manmade weapon)—together often known as natural and manmade EMP, or simply EMP—could cause a blackout lasting months or years. Even for government officials who have the authority to do something about it, legislation may be required to make new demands on a resistant, powerful industry.

For unfamiliar and intellectually intimidating topics, it may be necessary to educate legislators. The effort it takes to pass legislation to solve even relatively simple problems, however, may be enough to discourage legislators from voluntarily taking on this kind of new, unfamiliar challenge. Therefore, when facing the specter of a massive infrastructure problem and a powerful industry lobby, many default to a wait-and-see position.

Educating Legislators: Key sources of information for legislators are typically the legislation sponsor and supporters, the industry and its lobbyists, content experts, and outside interests, including the general public and the legislators' own supporters. The primary forum for educating legislators comes from a public hearing presented before the legislative committee that has jurisdiction over that policy area. Thus, to seek protections of the Maine electricity transmission system (the grid) from long-term blackouts due to GMD and EMP requires the public hearing to take place before the Energy, Utilities, and Technology (EUT) Committee.

As a state representative, it took a significant amount of time to learn about the threats of GMD and EMP, and to develop a substantial network of national experts on policy, science and technology, manufacturing, space weather, weapons, intelligence, and national defense. Dr. Peter Vincent Pry and the office of (now former) Congressman Roscoe Bartlett, both long-time national leaders on EMP, were significant in introducing politicians to experts who had been working on these issues at the federal level. Many of them came to Maine to testify at the hearing. These experts informed the EUT about threats to the electric grid that they had never heard about before from the power companies. They challenged the legislators to do the following:

- acknowledge that the State has a problem (as do all the states);

- recognize that the State has regulatory authority to fix the problem;

- identify available solutions and their costs (GMD protections exist that are low cost);

- provide effective leadership to protect Maine's electrical grid from long-term blackout; and

- serve as a model for other states.

The experts were articulate, convincing, and impressive when describing a compelling but scary message, so committee members were able to understand the issue.

On the other hand, the electric power industry "representatives" (lobbyists) who had spent careers lobbying for the industry before the EUT Committee (and other legislators) were not content experts, but rather public relations experts highly paid to deliver a message. They spoke positively about the electric companies' management of the threat, with statements including the following: "We are talking about a low-probability event; we have competing priorities; we've been protecting the grid for years; we are following all the NERC (North American Electric Reliability Corporation) reliability standards." Despite sounding impressive when delivering a reassuring message, they failed to answer key questions and to win over the committee. The threat they posed to passage of the bill was that they were familiar faces to the committee members—and their ingratiating smiles can tip the balance for lazy, confused, or just undecided legislators.

The Process Behind a Maine Bill: Facing news it could not ignore, the EUT lacked the confidence to act on or confront the industry's resistance, and amended the bill (LD 131, introduced by Andrea Boland) to a study, with the provision that the EUT could use its findings to draft permanent protective legislation the following year. The Maine Public Utilities Commission (PUC) was to conduct the study, and assured the EUT they could deliver it on schedule in January 2014. The industry agreed to the plan. LD 131 passed unanimously in committee as emergency legislation and in the House of Representatives, and passed by a vote of 32-3 in the Senate, to become law on June 11, 2013.

It was a deftly designed study and internationally acclaimed as model legislation. It also was the first ever EMP/GMD legislation passed in the nation. The Federal Energy Regulatory Commission (FERC) has an Office of Energy Infrastructure Security, which has a mission to assist states; its director, Joe McClelland, offered help with the study.

Two reports finally emerged—one influenced heavily by the electric power companies, and one supported by the independent experts—but not until 2015, and new elections had resulted in a newly configured legislature. Senator David Miramant introduced a new bill (LD 1363) to require installation of known, available protections supported in the studies. This time, the EUT split its vote, and the bill failed in the legislature—by one vote in the Maine Senate, along party lines. Low-cost solutions existed, and the prior legislature's nearly unanimous vote had supported emergency action to protect the grid, but the industry had succeeded in defeating it.

The difference in the results of the two legislative efforts may be explained by different factors at work. In 2013, the legislation, sponsor, and experts surprised the industry, which was unable to recover from the unexpected exposure of the threat and the apparent disinterest and/or incompetence of the power companies regarding GMD and EMP. In January 2014, the EUT chair, without a vote of the committee, had granted the PUC an extension to January 2015 to finish the study—under the direction of the biggest electric utility in Maine, Central Maine Power (CMP). By 2015, when LD 1363 was introduced, the industry had regained its political control, as the 2014 election had populated the EUT and one third of the full legislature with new faces. Various systemic political realities also may have contributed to the industry defeat of protections:

- Uneasiness about supporting a big, new, unfamiliar issue—It may not seem advantageous to some legislators to invest the time and effort to support a bill that might not pass, or to take a politically risky position opposing a political power industry. Legislative leaders remained quiet, not signaling support, maybe for similar reasons.

- Legislators' fears and lobbyist arguments, valid or not, to oppose the bill—Lobbyists make it easy for reluctant legislators to adopt their positions when they do not conduct their own research.

- Hesitation to cause trouble with big campaign donors—Legislative leaders are expected to raise money to get themselves and their members elected, and to fulfill an agenda.

- Committee chairs in Maine are appointed by legislative leadership (Speaker of the House and President of the Senate)—These leaders typically support the agenda of those who appointed them and often of the special interests under the committee's jurisdictions, and they are in a position to influence outcomes. The chairs never took up the PUC study reports for review, causing committee members to not be informed on their contents. Thus, they influenced the committee vote, which in turn, influenced the full legislature's vote.

- Appointment of committee members by leadership—Only three of the 2013 members of the EUT Committee were reappointed to the 2015 committee; 10 were new, including the chairs. Therefore, the committee did not benefit from a lot of experience with the subject.

- Influence of committee chairs—In 2013, the chairs did not limit the time visiting experts had to testify. In 2015, chairs limited them to three minutes each (meanwhile, the lobbyists were working every day in committee and in the halls of the State House). With so little input from the independent, national experts, and deliberately confused by lobbyists protecting electric companies from higher standards, new members were frustrated, unable to master critical new information, and split the committee vote. They thereby weakened the message to the rest of the legislature.

- The Senate chair of the EUT, Senator David Woodsome, who had been supporting the bill all along, changed his vote in the end, probably, as a new legislator, succumbing to party pressure, and spoke against it on the floor of the Senate. This was enough to defeat the bill by one vote, even though Senator Miramant spoke strongly for it. The House of Representatives had passed it decisively, where the three veteran EUT committee members spoke in favor of it.

Future Legislative Concerns: Many legislators who are motivated to follow and be politically safe, rather than lead on tough issues, often go along with party leadership or powerful interests. The legislative hierarchy structure, campaign funding laws, and committee system can work symbiotically to marshal votes for a separate agenda. Legislators who take on serious problems may find themselves opposing powerful interests and getting little or no help from their leadership because high political costs could reflect on them personally. Their constituents and the public in general may be strongly supportive, but not enough of them raise their voices.

Not unlike other powerful industries, the electric power industry uses media and lobbyists to telegraph an image of integrity and professional authority, but then uses inaccurate data in their studies to try to prove invalid arguments that work for them. To inexperienced, often stressed legislators, it may be persuasive. NERC, the electric power industry's association and lobbying arm, has sole authority to write its own "reliability standards" that determine their level of public responsibility. The Federal Energy Regulatory Commission (FERC) is charged with regulating NERC, but often turns to NERC for answers. In the same way, the Maine PUC turned to Central Maine Power Company for the LD131 study. CMP then turned to NERC, which provided data from another country, rather than using the Maine data it had, to support the outcome it wanted: the argument against GMD/EMP protections.

First-Hand Experience in the Maine Legislature: Big money and special interests have outside influence on the legislative process. It can often compromise leaders, defeat good legislation, endanger the public, and promote regulatory capture. It is difficult to display political courage when lobbyists of powerful interests smile and create confusion about the facts. For these reasons, testimony from subject matter experts needs to be treated with great respect. In this case, the testimony of first responders was very important. The public is critically important, too. Without public support, the nation cannot expect to maintain a self-governance.

The United States is the most vulnerable country in the world to natural and manmade solar storms and EMP because of its huge, interconnected grid and its dependence on electric power and electronics. State Senator and Navy veteran Robert "Bob" Hall of Texas refers to obstruction of protections of the grid as "treason" because it is also a national defense threat. Imagine what the fifth week of a blackout would be like following an EMP or solar storm: no heating, cooling, communications, water and waste systems, banking, hospitals, transportation, food delivery, etc.

Governing bodies must take charge of protecting the nation. If Congress is too conflicted to act, the states must. Many states are initiating action, but it is a struggle. In all states, the electric companies have blanket liability protection against the costs of catastrophe from these threats, so they have no incentive to act on their own to raise standards. The public must engage more and insist on more courage and dedication by their elected representatives, and more accountability from the electric power companies. They must be made to quickly repair the electric grid to a level of realistic protection against such horrific threats, and be held legally responsible to share in the consequences and real costs of catastrophes that result from their inaction.

Right now, the nation is in another pre-Katrina or pre-9-11 moment. A small army of people is working very hard to save the electric grid, and protect the nation, but it will take many more recruits, and bigger armies, moving governments, media, and industry in more states and in Washington DC, to win the war and save the country from the societal collapse that a severe GMD or EMP would threaten.

State Representative Andrea Boland is completing her eighth year in the Maine legislature. She is considered a leader in safety issues of electromagnetic radiation, especially from cellphones and smart meters. She became involved in electric grid protection against electromagnetic pulse and geomagnetic solar storms (GMD) at the suggestion of her regular scientific advisor. Her work is supported by several national experts. She earned a BA degree from Elmira College and an MBA from Northeastern University, and studied at the Sorbonne and Institute of Political Studies in Paris. She was awarded the 2011 Health Freedom Hero Award by the National Health Federation for her work on health freedom and safety. Her legislative work has led to confronting major corporate interests on matters of transparency and regulatory capture, and public protections.

Electromagnetic Pulses—Six Common Misconceptions

by GEORGE H. BAKER

Wednesday, November 5, 2014

Many misconceptions about electromagnetic pulse (EMP) effects have circulated for years among technical and policy experts, in press reports, on preparedness websites, and even in technical journals. Because many aspects of EMP-generation physics and its effects are obscure, misconceptions from those who do not perceive the seriousness of the effects to those who predict a doomsday chain of events are inevitable. However, not all EMPs are the same, with the most significant effects being caused by E1 and E3 fields.

Nuclear bursts detonated at altitudes above 40 km generate two principal types of EMPs that can debilitate critical infrastructure systems over large regions:

- The first, a "fast-pulse" EMP field, also referred to as E1, is created by gamma ray interaction with stratospheric air molecules. The resulting electric field peaks at tens of kilovolts per meter in a few nanoseconds, and lasts a few hundred nanoseconds. E1's broadband power spectrum (frequency content from DC to 1 GHz) enables it to couple to electrical and electronic systems in general, regardless of the length of their cables and antenna lines. Induced currents range into the thousands of amperes and exposed systems may be upset or permanently damaged.

- The second "slow-pulse" phenomenon, is referred to as magnetohydrodynamic (MHD) EMP, or E3, and is caused by the distortion of Earth's magnetic field lines due to the expanding nuclear fireball and the rising of heated, ionized layers of the ionosphere. The change of the magnetic field at the Earth's surface induces a field in the tens of volts per kilometer, which, in turn, induces low-frequency

221

currents of hundreds to thousands of amperes in long conducting lines only (a few kilometers or longer) that damage components of long-line systems, including the electric power grid and long-haul communication and data networks.

By over- and under-emphasizing realistic consequences of EMPs, policymakers may delay actions or dismiss arguments altogether. The six misconceptions about EMPs that are perhaps the most harmful involve: (a) exposed electronic systems; (b) critical infrastructure systems; (c) nuclear weapons; (d) cost of protection; (e) type of EMPs; and (f) fiber-optic networks.

Misconception 1: EMP Will Cause Every Exposed Electronic System to Cease Functioning. Based on the U.S. Department of Defense (DOD) and Congressional EMP Commission's EMP test databases, small, self-contained systems, such as motor vehicles, hand-held radios, and unconnected portable generators, tend not to be affected by EMPs. If there is an effect on these systems, it is often temporary upset rather than component burnout.

On the other hand, threat-level EMP testing also reveals that systems connected to power lines are highly vulnerable to component damage requiring repair or replacement. Because the strength of EMP fields is measured in volts per meter, the longer the conducting line, the more EMP energy will be coupled into the system, and the higher the probability of damage. As such, the electric power-grid network and landline communication systems are almost certain to experience component damage when exposed to an EMP with cascading effects to most other (dependent) infrastructure systems.

Misconception 2: EMP Effects Will Have Limited, Easily Recoverable, "Nuisance" Effects on Critical Infrastructure Systems. Although an EMP would not affect every system, widespread failure of a significant fraction of electrical and electronic systems will cause large-scale cascading failures of critical infrastructure networks because interdependencies among affected and unaffected systems. Mathematician Paul Erdos's "small-world" network theory applies, which refers to most nodes—equipment attached to a network—being accessible to all others through just a few connections. The fraction of all nodes changes suddenly when the average number of links per single network connection exceeds one. For example, a single component failure, where the average links per node is two, can affect approximately half of the remaining "untouched" network nodes.

For many systems, especially unmanned systems, loss of control is tantamount to permanent damage, in some cases causing machinery to self-destruct. Examples include the following:

- lockup, or not being able to change the "on" or "off" state, of long-haul communication repeaters;

- loss of remote pipeline pressure control in supervisory control and data acquisition (SCADA) systems, which communicate with remote equipment;

- loss of generator controls in electric power plants; and

- loss of machine process controllers in manufacturing plants.

Misconception 3: Megaton-Class Nuclear Weapons Are Required to Cause Serious EMP Effects. Due to a limiting atmospheric saturation effect in the EMP-generation process, low-yield weapons produce a peak E1 field similar in magnitude to high-yield weapons if they are detonated at altitudes of 50–80 km. The advantage of high-yield weapons is that their range on the ground is affected less significantly when detonated at higher altitudes.

Nuclear weapons with yields ranging from 3 kilotons to 3 megatons (a third order of magnitude difference in yield), when detonated at their optimum burst altitudes, exhibit a range of peak E1 fields on the ground differing by only a factor of ~3, viz. 15–50 kV/m. With respect to the late-time (E3, or low-amplitude, low-frequency components) EMP field, a 30-kt nuclear weapon above 100 km would cause geomagnetic disturbances as large as solar superstorms, although over smaller regions. It also is worth noting that peak currents on long overhead lines induced by E1 from 10 kiloton-class weapons can range in kiloamperes with voltages reaching into the hundreds of kilovolts.

Misconception 4: Protecting the Critical National Infrastructure Would Be Cost Prohibitive. Of the 14 critical infrastructure sectors, EMP risk is highest for electric power grids and telecommunication grids because of their network connections and criticality to the operation and recovery of other critical infrastructure sectors. Attention to hardening these infrastructure grids alone would provide significant benefits to national resilience.

The electric power grid is essential for sustaining population "life-support" services. However, some major grid components could take months, or years, to replace if many components are damaged. The primary example is high-voltage transformers, which can irreparably fail during major solar storms and are thus likely to fail during an EMP event. Protection of these large transformers would reduce the time required to restore the grid and restore the necessary services it enables.

According to Emprimus, a manufacturer of transformer protection devices, the unit cost for high-voltage transformer protection is estimated to be $250,000, with the total number of susceptible large, high-voltage units ranging from 300 to 3,000, according to Oak Ridge National Laboratory. The requirement and cost for generator facility protection are still undetermined but are likely to be similar to transformer protection costs. To protect SCADA systems, replacement parts are readily available and repairs are relatively uncomplicated. Protection costs for heavy-duty grid components are in the $10 billion range, which is a small fraction of the value of losses should they fail. When amortized, protection costs to consumers amount to pennies per month.

Misconception 5: Only Late-Time EMP (E3) Will Damage Electric Power-Grid Transformers. Oak Ridge National Laboratory's January 2010 report on its E1 tests of 7.2-kV distribution transformers produced permanent damage to transformer windings in seven of the 20 units tested. The failures were due to transformer winding damage caused by electrical breakdown across internal wire insulation. As an important side note, transformers with direct-mounted lightning surge arrestors were not damaged during the tests. Similar tests of high-voltage transformers are needed.

Misconception 6: Fiber-Optic Networks Are Not Susceptible to EMP Effects. In general, fiber-optic networks are less susceptible than metallic line networks; however, fiber-optic multipoint line driver and receiver boxes, which are designed to protect against ground current, may fail in EMP environments. Long-haul telecommunication and regional Internet fiber-optic repeater amplifiers' power supplies are particularly vulnerable to EMP environments (Figure 1). Terrestrial fiber-optic cable repeater amplifier power is provided by the electric power grid and thus vulnerable to grid failure as well as to direct EMP/E1 effects.

Undersea cable repeater amplifiers also are vulnerable to EMP/E3 effects since they are connected to a coaxial metallic power conductor that runs the length of the line. Because of its low-frequency content, E3 penetrates to great ocean depths, which subjects undersea power amplifiers to high risk of burnout. On the positive side, line drivers/receivers and repeater amplifiers are relatively easy to protect using shielding, shield-penetration treatment, and power-line filters and/or breakers.

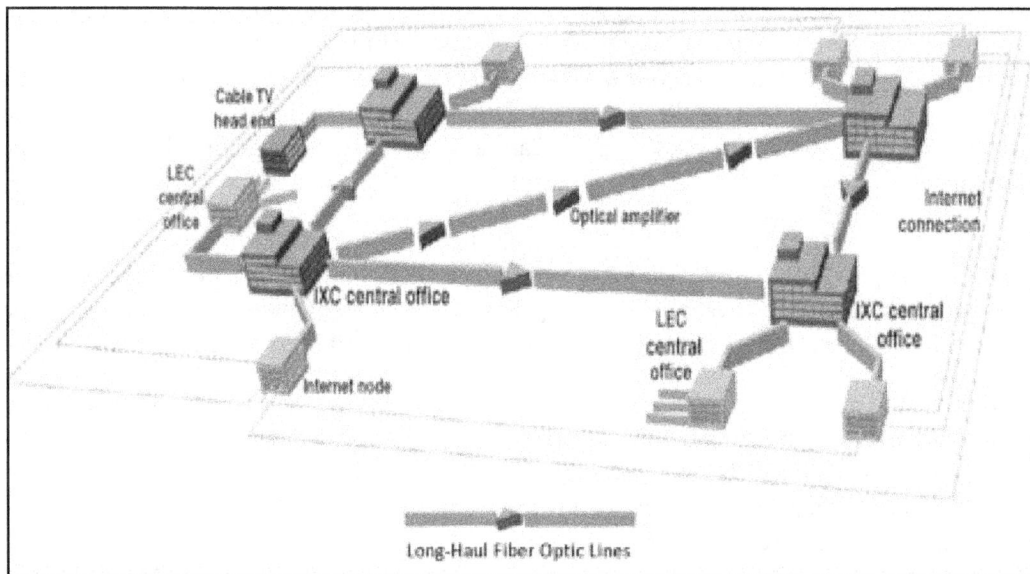

Long-Haul Fiber Optic Lines

Standardized Solutions: From a risk-based priority standpoint, the electric power grid is a high priority for EMP protection. Hardening this infrastructure alone would have major benefits for national resilience—the ability to sustain, reconstitute, and restart critical services. EMP engineering solutions have been implemented and standardized by DoD since the 1960s and are well documented:

- MIL-STD-188-125-1—"DOD Interface Standard—High-Altitude Electromagnetic Pulse (HEMP) Protection for Ground-Based C4I Facilities Performing Critical, Time-Urgent Missions—Part 1—Fixed Facilities" (July 17, 1998);

- MIL-STD-188-125-2—"DOD Interface Standard—High-Altitude Electromagnetic Pulse (HEMP) Protection for Ground-Based C4I Facilities Performing Critical, Time-Urgent Missions—Part 1—Transportable Systems" (March 3, 1999); and

- MIL-HDBK-423—"Military Handbook—High-Altitude Electromagnetic Pulse (HEMP) Protection for Fixed and Transportable Ground-Based C4I Facilities Vol. 1—Fixed Facilities" (May 15, 1993).

With respect to the power grid, the installation of blocking devices in the neutral-to-ground conductors of large electric distribution transformers will significantly reduce the probability of damage from slow EMP/E3. Transformer protection against E1 overvoltages is achievable by installing common metal-oxide varistors (control elements in electrical circuits) on transformers from each phase to ground. Costs for protecting the power grid are small compared to the value of the systems and services at risk.

George H. Baker is professor emeritus at James Madison University (JMU) and directed JMU's Institute for Infrastructure and Information Assurance. He consults on critical infrastructure assurance, specializing in EMP, and other nuclear effects. He is the former director of the Defense Threat Reduction Agency's critical system assessment facility. He also led the EMP group at the Defense Nuclear Agency responsible for development of the DoD EMP standards. He served as principal staff on the Congressional EMP Commission and now serves on the board of directors of the Foundation for Resilient Societies and the Congressional Task Force on National and Homeland Security advisory board. He earned a PhD in engineering physics from the U.S. Air Force Institute of Technology.

From the "Selected Works of George H. Baker," James Madison University, January 1992

Operation DESERT STORM demonstrated the clear military advantage that was provided by our sophisticated electronic C4I and weapons systems. In addition, our offensive tactic of taking out the adversary's eyes and ears during the air war paid off, giving the U.S. decisive air superiority. High tech means so dominated the battlefield that the outcome of future conflicts could be decided by electronics attrition rather than human casualties.

Our DESERT STORM experience accentuates the importance of guaranteeing that U.S. electronic systems will not be disabled either deliberately or accidentally by electromagnetic effects. The electromagnetic threat landscape is highly complex. The already formidable list of environments (EMI, lighting, ESD, EMP, HERO, TEMPEST, EW, etc.) is lengthened by emerging threats from high-power microwave (HPM) and ultra-wide band (UWB) EM weapons. Many of these environments overlap in the frequency and amplitude of the electrical stresses they create. The large number of electromagnetic effects to be dealt with is more than matched by the proliferation of applicable guidelines and standards and handbooks, which are approximately 100 at last count. These documents exhibit redundancy in requirements and mitigation techniques. In some cases (e.g., grounding techniques) guidelines are inconsistent or incompatible. Associated with the guidelines is a multitude of separate test requirements and simulator facilities that overlap in the environments they produce.

These factors provide a strong motivation to simplify requirements by unifying guidelines for dealing with electromagnetic effects. Fortunately, a fundamental physical basis for unification exists which was explored by Sir Michael Faraday. Theoretically, it is possible to exclude any extraneous electromagnetic environment by the use of a single, well-designed metal barrier, or Faraday cage. This barrier consists of two basic features:

    (1) a continuous metal enclosure; and
    (2) treatments for any necessary enclosure penetrations.

Examples of penetration treatments are such things as electrical filters for penetrating cables/wires, circumferential bonds for penetrating metal pipes, and metal screens over windows.

The EMP community has developed a system-level protection approach which fully embraces the Faraday cage method. The protection approach was debated over a period of many years and was ultimately, at the Defense Nuclear Agency's request, adjudicated by the National Research Council. After a one year in-depth study, the NRC Select Committee recommended that "stress within the system be controlled through integral shielding and penetration control devices to well known values," and that great emphasis should be placed on "developing better and cheaper means for virtually complete and effective shielding of systems … to include strong emphasis on early use of standardized shielded boxes interconnected with optical fibers." The NRC Committee believed it to be "imperative to use a design strategy that is testable." In the case of fixed ground-based facilities, this low-risk approach is now codified in Military Standard Number 188-125 (MIL-STD-188-125).

The Defense Nuclear Agency's low-risk EMP hardening approach is also suitable as a conceptual framework for unified electromagnetic effects (EME) protection. The low-risk concept provides conservative protection via electromagnetic barriers and penetration control over the broad EMP spectrum (0–1,000 MHz). The engineering principles involved may be extended to mitigate electromagnetic effects in general, including the HPM and UWB threats as well as other conventional effects. For example, the EMP approach used on the Pershing ground system has been experimentally demonstrated to be effective against HPM and UWB environments well into the gigahertz regime.

Of necessity, the EMP approach is quite conservative due to the continental scale geographic coverage of EMP from a single weapon and the likelihood of EMP exploitation during periods of extreme national emergency when critical electronics are most needed. Understandably, not all system programs can afford such a conservative approach. However, because of the effectiveness of the Faraday barrier against multiple environments, the EMP approach is highly suitable as the starting point and conceptual basis for developing unified EME protection guidance. Elements of the EMP approach can be applied to greater or lesser degrees on a system-by-system basis depending on the EMEs of interest and the level of acceptable risk. The fundamental elements of a low-risk approach include the following:

1. Shielded enclosures for mission critical electronics (this shielding does not have to enclose the entire system).
2. Minimum number of penetrations through the shield.
3. Protected treatments for all penetrations, including conducting cables and breaches.
4. Designation of interface points for EMP stress and system strength comparisons.
5. System performance tests to threat relatable EM environments.
6. Life cycle surveillance and maintenance of protection features.

To explore avenues for unification, the Defense Nuclear Agency and the U.S. Army Nuclear and Chemical Agency jointly hosted a Unified EME Hardening Workshop in October 1991. We were greatly encouraged by the workshop proceedings. An executive panel comprised of representatives from the Services, the Defense Nuclear Agency, the Strategic Defense Initiative Office, and industry convened at the close of the workshop to discuss issues surrounding electromagnetic environments and effects (E3).

The panel reached a strong consensus and issued the following recommendations:

1. DoD's E3 community should define a common set of terminology.
2. The EMP community should provide a more general definition of what is meant by low-risk protection.
3. A top-level road map for E3 should be developed. This could take the form of an umbrella DoD instruction which explains when to invoke available guidance and standards.
4. An effort is needed to review existing hardening guides to resolve incompatibilities.
5. Hardening design guidelines/procedures/processes should be integrated to handle multiple (most if not all) E3 requirements.
6. DoD and the Services should establish policy, rationale, and procedures to incorporate "new" electromagnetic threats into system requirements.
7. The Services should collaborate in sharing their E3 lessons learned and consider establishing a joint database.
8. Future EM test facilities (new and consolidated) need to consider unified E3 issues.
9. An effort is needed to explore the feasibility of developing consolidated tests to replace individual threat tests.
10. The Defense Acquisition Board should impose a minimum set of reporting requirements covering stress reduction, strength determination, verification testing, and life cycle maintenance of protection (this could be satisfied by requiring an E3 program plan for systems).
11. A DoD/industry E3 working group should be established to implement these recommendations.

A main priority for this group should be recommending more descriptive E3 text for inclusion in DoD Instruction 5000.2. Now is the time for a concerted effort to unify guidance for treating all electromagnetic environments and effects, at the same time incorporating the emerging HPM and UWB threats. The DoD electromagnetic protection community needs to look for guidance and commonalities to unify our policy in order to streamline the process. An additional impetus for unifying EME protection is provided by the current OSD initiatives to consolidate laboratories. Unification will require: (1) top-level, uniform guidance for treating (E3) with the DoD 5000 series as a possible vehicle; (2) a road map to enable system program offices to screen the large number of existing guidelines, standards, and handbooks and, at the same time, consolidate material; and (3) the establishment of a DoD focal point for E3.

For additional information see:
National Research Council Report. 1984. "Evaluation of Methodologies for Estimating Vulnerability to Electromagnetic Pulse Effects."

This essay was part of a complete issue focused on EMP and extreme space weather that DomesticPreparedness. com published in November 2014 named Blackout. EMP SIG members contributed a number of articles that are being reprinted in these proceedings. That complete issue can be found here: http://www.domesticpreparedness. com/pub/docs/DPJNovember14.pdf

This article by Charles Manto can be found at this website address: http://www.domesticpreparedness.com/Infrastructure/Building_Protection/Electromagnetic_Pulse_Triage_%26_Recovery_/. This article is reprinted with permission from the IMR Group, Inc., publisher of DomesticPreparedness.com, the DPJ Weekly Brief, and the DomPrep Journal. The IMR, Inc. offers no guarantees as to the accuracy of any information presented, but encourages all readers to use IMR, Inc. programs primarily as a resource to facilitate their own research.

Electromagnetic Pulse Triage and Recovery

by CHARLES (CHUCK) L. MANTO
Tuesday, November 25, 2014

Emergency medical technicians and paramedics triage patients for mass-casualty incidents. Hospital emergency rooms have triage nurses to determine the level of care needed. Community Emergency Response Team (CERT) participants are taught how to quickly sort and tag victims so that they can focus on the seriously injured and sustain them until help arrives. Conversation about high-impact disasters should convey hope. Without the hope of recovery and the management capacity implied in the word "triage," the problem may seem overwhelming. One sign of hope is that an economic impact assessment on electromagnetic pulse (EMP) showed that protecting even 10 percent of the most critical infrastructure could avoid up to 85 percent of losses. Triage helps identify that 10 percent.

Triage is especially helpful in the case of a high-altitude nuclear burst EMP or severe solar storm. Depending on who is in charge and where the impacted organization is located relative to the event, actions will range from relatively easy to nearly impossible. Knowing the difference makes it possible to begin developing and deploying plans now in order to manage and recover from such incidents in the future.

How EMP Works: Large solar storms create ground-induced currents similar to the slow-rise time pulse of a large high-altitude nuclear EMP burst, known as E3. Currents can connect with conductors in the ground to damage equipment connected to ground wires, including large transformers that may take over a year to replace.

A high-altitude nuclear burst from even a small weapon could disrupt or damage electronics at nanosecond speeds within a specific region and, under the right circumstances, across the continental United States. Smaller electronic hand-held or vehicle-mounted electromagnetic interference (EMI) devices only act on equipment that is at fairly close range and require a number of people to cause interruption in a large area, but can pose serious threats like other coordinated physical attacks. Protecting equipment from high-altitude EMP also protects against smaller EMI weapons.

Manmade EMP creates both a radiated pulse through the air and a conducted pulse along cables. See the article by George Baker, professor emeritus at James Madison University, for more-detailed explanations and discussions of common misperceptions about EMP.

For any triage scenario, sometimes the most difficult tasks are ultimately "written off." For example, although protecting large power utility generators and transformers is relatively inexpensive, a moderately difficult activity for utilities would be nearly impossible for an outside firm to impose on these utilities. At this time, only a few utilities have begun the process of protecting their most critical assets and earning certification by independent testing authorities to prove they meet objective standards of protection from either EMP or a hundred-year solar storm. The InfraGard National EMP Special Interest Group's (SIG) conference proceedings of 2012 and 2013 include technical, economic, policy, and emergency management resources, which help corporate as well as local and state government officials to assess: (a) what can and is being done (or not done) at the federal and industry-wide levels; and (b) what, in this class of threats, requires an all-of-nation response.

Benefits of EMP-Protected Microgrids and Local Networks: From a triage perspective, scenarios involving a nationwide collapse of centralized infrastructures for 1–12 months or longer would highlight the need for companies and organizations to avoid total dependence on large centralized systems outside of their own control and plan. For this reason, the EMP SIG is creating a workshop and tabletop exercise package along with background technical and engineering information that it will make available to state and local government emergency management after the completion of its workshop and facilitated discussion on December 4, 2014.

As organizations face the prospective collapse of centralized infrastructure, they may become motivated to discover how to make and store enough power locally to continue to operate indefinitely without a centralized power grid and communications network. In fact, protected microgrids and local networks would make the centralized grids more resilient because those islands of sustainable power would minimize the domino effect of cascading failures inherent in a regional or nationwide incident.

Industries and communities that produce some of their own power would save or make money by avoiding peak load charges. Local communities that already produce their own power will find it easier to protect their assets than with larger systems. Fortunately, for those wanting to produce and store power for their own facility or campus, technologies are available and new technology in the near future will make it more cost competitive with centralized systems, especially given the resulting improved sustainable reliability. There are some companies including utilities that are offering systems integration services for microgrids. In time, they also could provide EMP-rated microgrids.

Local EMP Protection—The Easy Part of EMP Triage: Fortunately, organizations can take some relatively easy and inexpensive steps on their own. For example, since EMP and intentional EMI are either radiated through the air or conducted through power or communication lines, one easy and nearly free action any organization can take to reduce the probability of EMP disrupting or damaging critical equipment would be to unplug equipment that is not being used. Emergency operations centers, which are often reserved for emergencies and not used much of the time, have multiple computer and communication stations connected to power and communication lines 24/7. If unused subsystems were to be unplugged when not in use, then the conducted pulses would not be as likely to couple with the systems through those conductors.

Circuit breakers or plug connections could be deployed in easy-to-see locations at eye level so facility managers can walk into rooms and see at a glance whether unused systems are plugged in. Requiring managers to walk around and crawl under desks to see if something is plugged in is not a sustainable maintenance method. Unplugged systems still would be vulnerable to EMI through the air, but unplugging them when not used would increase the chances of survival. In addition, less power used for standby capability would actually save money for organizations. The savings then could be used to purchase EMP-rated surge protectors or electronic filters for electric distribution systems so those particular lines would be able to pass through power while filtering excess EMI during operation. Filtered lines also deliver "cleaner" power to devices, minimizing the impacts of small day-to-day surges and fluctuations that reduce equipment health and lifespan.

Perhaps the simplest and least expensive measure to protect equipment from radiated EMI would be to place spare equipment into EMP-shielded containers or rooms. In this case, solid metal containers that are independently EMP rated are the most reliable solutions. Homemade solutions also may be effective, especially for those who are required to have fire-rated steel safes for files, equipment, or firearms. In these cases, there should be no holes into the safe, and doors should have metal gaskets around all of the edges so signals do not travel through the seams or edges of the closed door.

Similarly, steel or aluminum garbage cans for this use should have metal tape applied to the seams and metal gaskets of the same or compatible metal as the can around the inside of the lid. Aluminum tape or gaskets applied to an iron or steel surface, for example, would result in corrosive interactions between the different metals, which would degrade the shielding. In addition, the inside needs to be lined with a nonconducting insulation layer to protect equipment from the metal layer that will hold or pass a charge.

Somewhat More Difficult Local EMP Protection—Moderate EMP Triage: Shielding operating equipment and rooms from airborne EMI is a little more complicated and expensive because these rooms have power, communications, and air circulation that make shielding more difficult but not impractical. Business arrangements prove that even cash-poor organizations among counties, hospitals, or universities can acquire equipment at no monetary cost by providing in-kind resources to business continuity parks that would provide protection to organizations as they create EMP and cyber-resilient local networks supported by local power generation and energy storage systems.

Some buildings constructed entirely of steel may have enough inherent shielding properties built in that could be modified to provide a small measure of EMP protection. For example, some all-steel buildings may have enough shielding value in their material that could provide as much as 30 decibels (dB) of the 30–100 dB of protection required by military specifications if those buildings were to be modified as a consistent shield. Even a lower level of shielding such as this will improve the odds that equipment might survive a given EMP event. (Every 20 dB of protection reduces the amount of EMI passing through the shield by a factor of 10.)

Meeting the Military Standard Tests: The military standard 188.125 for EMP protection currently requires a minimum reduction of the pulse by a factor of 1,000 or 80 dB—that is, the reduction deemed necessary to allow otherwise vulnerable equipment to operate without disruption while under an EMP attack. Even these levels of protection do not have to be cost-prohibitive, especially if built into infrastructure in the beginning of the planning process.

Operating equipment can be shielded from radiated pulses by placing equipment into cabinets or rooms that are shielded on all six sides by either welded (best) or bolted (next best) metal to ensure protection at a given level. These rooms and equipment also must be protected from conducted pulses since cables either act as antennae that promulgate the pulse into the room or provide direct pathways into the equipment they connect. All power and communication wires must be filtered from excess electromagnetic energy. Conductors of any sort must be placed into a shielded space with proper filtering and connections.

Figure 1: An EMP-rated waveguide for airflow.

In addition, waveguides can protect air passageways by capturing the electromagnetic waves and connecting them back to ground connections (see Figure 1). EMP-rated doors and sally ports need to be tested to ensure that they do not compromise the rooms (see Figure 2). Ideally, management will support and determine that continued maintenance and testing will be performed on the most critical infrastructure deserving this level of protection.

Figure 2: An EMP-protected door and filter system as part of an EMP protected 8x20 foot cargo container.

It is critical that subsequent changes do not damage the shielding effectiveness. Business continuity, security, change management (i.e., a management process used to make systems, usually information technology and infrastructure, to ensure consistency with overall requirements), and maintenance systems should all be integrated to ensure the alignment of management objectives and day-to-day practice. More than once, multimillion dollar facilities have been compromised because someone decided to casually drill a hole through a shielded room to place antennas through the shield for a better communications signal. Experienced contractors could ensure that the work is tested against objective standards and the desired.

EMP protection is maintained over time.

Simplifying a Civilian Critical Infrastructure EMP Rating System: Civilian critical infrastructure managers would be better served by a simple and understandable EMP rating system such as the one used by the Uptime Institute for data centers, which ranks them from low (Level 1) to high (Level 4) based on their resilience. One industry practice proposed by Instant Access Networks takes a similar approach by providing objective standards against which EMP-protected facilities and equipment could be measured. In this approach, Level 3 would be a way to meet all harmonized military standards for EMP and Tempest (signal emitting protection) at a 100-dB level. For those who want a greater level of protection, Level 4 provides 140 dB of protection, while Level 1 provides 30 dB and Level 2 provides 60 dB.

This is especially useful in a systems approach to civilian critical infrastructure that can be composed of rooms, systems, facilities, campuses, and networks spanning regions or continents. Not all components or systems will require the same amount of protection and not every system is equally valuable. An overall design approach to business continuity will require different levels of protection depending on the importance of the system element being protected and its vulnerability. Instant Access Networks LLC devised the following way to show ranges of protection that may be relevant to various elements—power, data, and communications—in a four-level, system-wide protection method where Level 3 meets or exceeds various military specifications for EMP and Tempest requirements.

Large data center and control rooms have been constructed to meet EMP standards for the military for decades. In the past year, a U.S.-based utility created an EMP-protected control room in Texas and an insurance company built an EMP-protected data center in Pennsylvania. Mass producible EMP-protected equipment such as cabinets, transportable 8×20 foot cargo containers, mobile command centers, and microgrids could be deployed across networks and control facilities so racks of computer or communications equipment and their power systems could be protected within them.

Getting Help: EMI threats can be complicated to understand and even tougher to mitigate. However, by following these simple triage steps, organizations can acquire EMP protection that can also protect life-sustaining systems from the effects of smaller EMI weapons and larger solar storms. In order to confidently take steps to protect critical infrastructure, here are three categories of available resources:

- *Written resources:* In addition to the conference proceedings of the InfraGard National EMP SIG, there are a number of information sources. First, there are general studies such as the work of the U.S. Congressional EMP Commission and various public domain military standards and manuals. Second, there are organizations such as the IEEE Electromagnetic Compatibility Society (EMC-S) that looks at various types of EMI that can be caused either intentionally or accidentally. The EMC-S also holds regional meetings and an annual meeting around the country. Trade magazines such as Interference Technology will cover issues that include intentional EMI and covers vendors that are active in the field.

- *Corporate resources:* Some interesting and helpful resources are testing companies and organizations, which understand the standards and how to meet them. It is possible to develop relationships with testing company staff well enough to ask what to look for when evaluating vendors. These include companies that test systems and buildings for EMP protection, such as SARA, Inc. or Jaxon Engineering in Colorado Springs, Colorado, and Little Mountain Test Facility at Hill Air Force Base in Ogden, Utah.

- *Conferences and tabletop exercises:* The InfraGard National EMP SIG is conducting its annual conference at the Dupont Summit on December 5, 2104 and will hold a by-invitation-only tabletop exercise on December 4.

When purchasing products that claim to be EMP protected, it makes sense to ask the vendor: (a) what they mean by that; (b) which objective measure they are using; and (c) whether the product has been tested by an independent testing organization. It also is important to know when a company claims that their product meets an EMP testing standard to know if it meets the entire standard or just part of it. There are some who are experienced in offering EMP triage consulting that also can be engaged. These simple precautions will be helpful as the urgency, importance, and affordability of EMP protection becomes apparent both on its own merits and as part of a total cyber-security framework. Working together with other users also will make the entire process more interesting, effective, and affordable.

Charles "Chuck" Manto is chief executive officer of Instant Access Networks LLC, a consulting and research and development firm that produces independently tested solutions for EMP-protected microgrids and equipment shelters for telecommunications networks and data centers. He received six patents in telecommunications, computer mass storage, and EMP protection and has another one pending for a smart microgrid controller. He is a senior member of the IEEE and founded and leads InfraGard National's EMP SIG. He can be reached at cmanto@stop-EMP.com.

This article is adapted from Ambassador Cooper's October 22, 2014 presentation at a South Carolina InfraGard Members Alliance Conference on Sullivan's Island. Ambassador Cooper's instructive historic perspective can be summarized in his own words: "When President John F. Kennedy announced that Soviet ships were transporting nuclear weapons and ballistic missiles to Cuba, U.S. citizens prepared to 'duck and cover' as they had been taught in grade school. Individuals and families were more self-reliant in the 1960s than today. With greater reliance on electricity, all Americans are now even more vulnerable, especially to the electromagnetic pulse (EMP) from a high-altitude nuclear burst. Loss of the electric grid now would freeze the United States' 'just-in-time' economy, leaving most Americans without means for survival—they could die within a year."

Challenge: Defeat Ballistic Missile Attacks From the South

By Henry F. Cooper

October 22 marks the 53rd anniversary of President John F. Kennedy's television announcement that Soviet ships were transporting nuclear weapons and ballistic missiles to Cuba—constituting a mortal threat to the United States. Miami was only 90 miles away and those missiles could have reached much farther.

Actually, the situation was worse than President Kennedy knew. After the Cold War, former Soviet authorities revealed that 100 nuclear weapons were already in Cuba and Fidel Castro wanted to keep them. Soviet General Secretary Nikita S. Khrushchev overruled Castro and removed them as the crisis wound down. Notably, the United States removed its nuclear-armed missiles from Turkey.

Many believe this crisis was the closest America came to nuclear war during the Cold War. The United States now confronts another existential threat from the South, in at least two ways.

A Modern Nuclear Missile Crisis

As illustrated in Figure 1, a North Korean freighter was caught in June 2013 smuggling military cargo from Cuba through the Panama Canal, into the Gulf of Mexico. Included were two SA-2 missiles, each capable of carrying a nuclear weapon, as the Soviets designed during the Cold War. They could have been launched to attack the United States from that freighter—or Cuba, Venezuela, or some other country—and currently there is little or no defense against them.

Today, we are on the brink of another threat "from the South"... To which we seem to be just as oblivious as we were in 1962!

- *A Wake Up Call: June 2013 intercept of a North Korean ship carrying from Cuba to & through the Panama Canal nuclear capable SA-2s and other technology illustrates the "Cacophony of Proliferation"*
- *Of greater concern, Iranian (or terrorist) missiles could be launched from ships off our coasts, especially in the Gulf of Mexico and/or from Latin America, e.g., Venezuela*
- *North Korean or Iranian Satellites could carry nukes over the South Pole to attack the U.S.*

Iran has launched satellites into orbit to the South, over the South Pole

North Korea also launches to the South

We are currently defenseless against these threats from the South!

Such nuclear weapons need not be exploded in U.S. cities. Indeed, their effects would be far more devastating if detonated at high altitude to produce an electromagnetic pulse (EMP) to debilitate major segments of the currently unhardened U.S. electric power grid—with cascading disastrous effects over much of the nation.

This threat was identified by the congressionally mandated EMP Commission in 2004 and 2008 and reiterated by many informed authorities, but the grid is still unhardened against such attacks. In addition, the United States operates no missile defense against such threats, although the U.S. government could rapidly and inexpensively deploy one.

Figure 1 also illustrates that Iran and North Korea launch satellites southward such that they pass over the South Polar region and approach the United States from the South. North Korea announced: (1) on September 14, 2015, that it had restarted the Yongbyon nuclear reactor and was ready to use nuclear weapons "any time" against the United States, and (2) the next day that it was preparing to launch a satellite (no doubt with Iranian scientists and engineers present as in the past). There is no reason to doubt that this threat is real—and that it provides a pathway for Iran to "outsource" its nuclear, ballistic missile and space launch development programs.

If such a satellite carries a nuclear weapon, it can avoid our Northward looking Ballistic Missile Defense (BMD) systems and detonate that warhead above the United States in its first orbit to produce an EMP that could take down the unhardened U.S. electric grid for an indefinite period. This attack would return America's current just-in-time economy to that of the nineteenth century and leave most Americans without the life support and security of the agrarian society of that era.

Credible estimates suggest most of the over 300 million people in the United States would die within the next year from starvation and the consequent chaos. For example, Dr. William R. Graham—former director of the White House Office of Science and Technology Policy and chairman of the EMP Commission—so testified before the House Armed Services Committee on July 8, 2008, and numerous others echo his and the commission's statements.

Response: Upgrade U.S. Missile Defenses

In addition to hardening the electric grid—which will take substantial time, we need much more effective BMD systems than the ground-based sites in Alaska and California that provide limited defenses against North Korean and Iranian intercontinental ballistic missiles (ICBMs) that attack the United States from the North. The government plans to deploy another ground-based site in the northeastern United States to increase that capability, especially against Iranian ICBMs.

However, the threat from over the South Polar region (and from off our east and west coasts) has largely been ignored. In particular, Americans are defenseless against ballistic missiles launched from vessels in the Gulf of Mexico or from a "satellite" attack from the South—patterned after the Soviet Fractional Orbital Bombardment System (FOBS) of the Cold War.

To counter ballistic missiles launched from offshore vessels, the United States could employ its sea-based defenses, deployed on nearly 35 cruisers and destroyers at sea around the world. If prepared to engage, such defenses are inherently capable of shooting down missiles launched from almost anywhere.

The U.S. Navy has repeatedly demonstrated that the Aegis BMD system is capable of shooting down attacking missiles while they are ascending from their launch points and above the Earth's atmosphere on their way to their targets. However, the ships must be appropriately located, with crews trained and ready, to shoot down missiles intended to attack the United States.

On a random day in 2013, there were four to six Aegis BMD ships along the eastern seaboard or in an east coast port. Under this condition, the east coast can be defended if the Aegis BMD ships are appropriately located and their crews are trained and ready to engage. Those in charge of homeland defenses can assure these conditions are met.

However, Aegis ships usually do not traverse the Gulf of Mexico, so threats from the Gulf likely would remain. The United States could affordably purchase and deploy on military bases around the Gulf the same Aegis Ashore system it is building in Romania (operational by year's end) and Poland (to be operational in 2018). No new development is required, just that the citizens and their local and state authorities agree to place these systems on bases such as Tyndall Air Force Base in the Florida panhandle—the location of First Air Force, the command responsible for the air defense of the continental United States.

This country is building Aegis Ashore bases to protect NATO allies against Iranian ballistic missiles; surely it can build the same "football-size" installations to protect U.S. citizens. There already is a site for testing in Hawaii. Sites around the Gulf of Mexico, and possibly along other coasts, could close operations gaps in defense coverage provided by the normally operating Aegis BMD ships.

Furthermore, the first generation Aegis system was used in 2008 to shoot down a dying satellite before it could spread its toxic waste on populated regions on Earth. Today's improved Aegis BMD system retains an improved inherent capability to support homeland defense missions involving threats from satellites approaching North America from the South. What is required technically is sensor information to launch the Aegis interceptors on the right track to complete the intercept with their on-board sensors—a continuously improving global sensor capability.

If a satellite is at an altitude that exceeds the Aegis interceptor's altitude range, then ground-based interceptors on Vandenberg Air Force Base, California, can complete the intercept, provided they have the needed upstream sensor cues. This capability can be provided by forward operating Aegis BMD ships, ground-based radar, and space-based sensor systems. For example, a ground-based radar in the Philippines could cue the Vandenberg BMD system.

Another alternative is to snuggle an Aegis Cruiser near the North Korean satellite launch pad and shoot the satellite launcher down as it is launched.

To be effective in either case, a political call is required, directing the military to shoot down the satellite is a timely way. It would be most helpful if the President made a unilateral declaration that this would be our policy, unless the satellite was proven not to constitute a threat by means of appropriate inspections of the satellite payload—e.g., with radiation detectors that would not compromise North Korean design information.

Bottom Lines

Other initiatives to help counter the existential EMP threat were discussed in an October 3, 2014 *Investor's Business Daily* article and elaborated in High Frontier's email message on October 8, 2014—further elaborated in subsequent High Frontier messages—see www.highfrontier.org . The above discussion simply emphasizes the reality of the manmade, ballistic missile threat and that there are possible, essentially off-the-shelf, means to defend against it.

The electric grid also should be hardened against nuclear EMP effects because no defense is perfect. If that is done, the grid also will be hardened against nature's EMP threat, which is produced by solar storms. No defense will protect against natural EMP from a solar storm that will one day occur, but defenses can be built faster than the grid can be hardened. Both remedies to the nation's current vulnerability should be initiated without further delay.

Finally, hardening the grid against only the solar storm-produced EMP will not assure the grid is hardened to the shorter wavelength EMP threat posed by a high-altitude nuclear explosion. If, for political reasons, the solar threat is taken as the primary reason for hardening the grid, that hardening effort should accommodate the entire EMP spectrum to protect against nuclear EMP attack.

Ambassador Henry (Hank) F. Cooper is chairman of High Frontier and a former acquisition executive for all U.S. ballistic missile defenses. He also served in several other senior U.S. government acquisition and policy positions, including as President Ronald Reagan's chief negotiator at the Geneva Defense and Space Talks with the Soviet Union and U.S. Air Force deputy assistant secretary for strategic and space systems. He currently is focused on helping local, state, and federal authorities protect against the natural and manmade electromagnetic pulse threat by building effective ballistic missile defenses and hardening the electric grid. This article is adapted from his October 22, 2014 presentation at a South Carolina InfraGard Members Alliance Conference on Sullivan's Island.

The Role of the Critical Infrastructure Protection and Recovery (CIPR) Working Group within the International Council on Systems Engineering (INCOSE)

The **International Council on Systems Engineering** (INCOSE) is a not-for-profit membership organization founded to develop and disseminate the interdisciplinary principles and practices that enable the realization of successful systems. INCOSE's mission is to share, promote, and advance the best of systems engineering from across the globe for the benefit of humanity and the planet.

INCOSE has grown significantly since its formation in 1990 with a membership that represents a broad spectrum—from student to senior practitioner, from technical engineer to program and corporate management, and from science and engineering to business development. Members work together to advance their technical knowledge, exchange ideas with colleagues, and collaborate to advance systems engineering. We currently have nearly 10,000 members in 62 countries.

INCOSE currently has 46 working groups that address many focused systems engineering topics (e.g., requirements, architecture, model-based systems engineering, anti-terrorism international, system of systems, complex systems, infrastructure, transportation, life cycle management, standards, healthcare, risk, space, security, complex systems, resilient systems, reliability engineering, tools). All WGs are focused on understanding their respective topics and improving their systems engineering applications, processes, etc. Goals:

- To provide a focal point for dissemination of systems engineering knowledge.
- To promote international collaboration in systems engineering practice, education, and research.
- To assure the establishment of competitive, scale-able professional standards in the practice of systems engineering.
- To improve the professional status of all persons engaged in the practice of systems engineering.
- To encourage governmental and industrial support for research and educational programs that will improve the systems engineering process and its practice.

Systems engineering is an interdisciplinary approach and means to enable the realization of successful systems. It focuses on defining customer needs and required capability early in the development cycle, documenting requirements, then proceeding with design synthesis and system validation while considering the complete problem. Systems engineering integrates participating disciplines and specialty groups into a team effort by coordinating contributions throughout the system life cycle stages from concept to disposal. Systems engineering balances the social, business, and technical needs of all stakeholders to achieve a quality product that meets these needs.

With our 25-year history supporting industry, INCOSE is able to inspire and guide the direction of systems engineering and envision the future state as this century continues to unfold. We are committed to shaping a future where systems approaches are preferred and valued in solving problems, whether enabling holistic solutions to global challenges or providing solutions for product development issues. Vision 2025 is a publication produced by a team of leaders from industry, academia, and government with the intent that it will be used by people working in many domains—healthcare, utilities, transportation, defense, finance—who will add their unique perspectives to the role systems engineering plays in our serving our world's many complicated demands. Vision 2025 addresses:

- The Global Context for Systems Engineering
- The Current State of Systems Engineering
- The Future State of Systems Engineering.

Participate in the future—work with INCOSE to bring forward your domain perspective.

Critical Infrastructure Protection and Recovery (CIPR) Working Group

INCOSE seeks to answer the call for help to resolve complex systems issues of national and international importance. When approached by InfraGard for help, we chartered the Critical Infrastructure Protection and Recovery (CIPR) Working Group in June 2015. The purpose for the CIPR Working Group (WG) is to provide a forum for the application, development, and dissemination of systems engineering principles, practices, and solutions relating to critical infrastructure protection and recovery against manmade and natural events causing physical infrastructure system disruption for periods of a month or more.

This WG will provide and support opportunities to exchange knowledge and systems engineering information and solutions within the scope of the CIPR WG, both within INCOSE and with external organizations sharing similar interests and goals. The opportunities include systems engineering products (e.g., architectures, requirements, IV&V, etc.). This information will be disseminated through publications (papers, articles, briefings) and supporting meetings, conferences, panels, and other means.

Specific areas of knowledge include the following:

a. The events capable of causing infrastructure disruption for periods of a month or more, to include all aspects of their characteristics and impacts.
b. The socio-technical factors related to CIPR.
c. The overarching structure and inter-connectedness among the critical infrastructure domains.
d. The interaction among infrastructure systems under various degraded states of operation.
e. Possible conceptual and design solutions, and related information.
f. Strategies for verification and validation of solutions.

Certain manmade and natural events have a known potential to affect societies at a national, continental, or even global scale. Such events can cause extreme harm well beyond those experienced from regional catastrophic events, especially when the effects will take longer than a month to recover. Three examples of events with the potential to cause critical infrastructure collapse include solar storms caused by coronal mass ejections (CME), electromagnetic pulse (EMP), and cyber events (intentional and otherwise). The CIPR WG will pursue its goals by addressing these three classes of events, and other classes of events with similar potential, when identified. The CIPR WG will promote and apply systems engineering principles with emphasis on policy, analysis, and concepts useful to understand, protect, and recover existing operational infrastructure, and to provide strategies, standards, and concepts for more resilient approaches. It will promote and perform activities supporting the stated goals.

The critical infrastructure domains addressed by the CIPR WG include the following. Other domains may be addressed as the need is identified. Chemical and other industrial bases; Communications, electrical & energy production and distribution; Emergency services; Financial services; Food and agriculture; Government services and facilities; Healthcare and public health; Information technology; Nuclear reactors, materials, and waste; Transportation; Water storage, treatment, and distribution; Waste handling and disposal (water, refuse, hazardous); Society at large.

The CIPR working group is seeking new members for the United States and international groups. We are also seeking collaboration with individuals and groups external to INCOSE with a passion to see the CIPR solutions advance for the sake of civilization. The CIPR WG will endeavor to integrate among governmental, business, and industry organizations and societies to define and promote strategies and solutions, while helping preserve intellectual property and sensitive information from improper disclosure.

To participate with the CIPR WG and for more information, please contact the following chairs:
Mike deLamare   madelama@bechtel.com
Loren (Mark) Walker loren.walker@bct-llc.com
John Juhasz   telepath.juhasz@yahoo.com

Initial Charter of the INCOSE Critical Infrastructure Protection and Recovery (CIPR) Working Group (WG)

1 PURPOSE

The purpose for the is to provide a forum for the application, development, and dissemination of systems engineering principles, practices, and solutions relating to critical infrastructure protection and recovery against manmade and natural events causing physical infrastructure system disruption for periods of a month or more.

Critical infrastructures provide essential services underpinning modern societies. These infrastructures are networks forming a tightly coupled complex system cutting across multiple domains. They affect one another even if not physically connected. They are vulnerable to manmade and natural events that can cause disruption for extended periods, resulting in societal disruptions and loss of life.

The inability of critical infrastructures to withstand and recover from catastrophic events is a well-documented global issue. This is a complex systems problem needing immediate coordinated attention across traditional domain and governmental boundaries. For example, the U.S. President issued Presidential Policy Directive PPD-21 that addresses "a national unity of effort to strengthen and maintain secure, functioning, and resilient critical infrastructure." This includes an imperative to "implement an integration and analysis function to inform planning and operations decisions regarding critical infrastructure." This working group will seek to support this and other policies with international reach.

INCOSE, as the premier professional society for systems engineering, can provide significant contributions toward critical infrastructure protection and recovery.

2 GOALS

This WG will provide and support opportunities to exchange knowledge and systems engineering information and solutions within the scope of the CIPR WG, both within INCOSE and with external organizations sharing similar interests and goals. The opportunities include systems engineering products (e.g., architectures, requirements, IV&V, etc.). This information will be disseminated through publications (papers, articles, briefings) and supporting meetings, conferences, panels, and other means.

Specific areas of knowledge include the following:

a. The events capable of causing infrastructure disruption for periods of a month or more, to include all aspects of their characteristics and impacts.
b. The socio-technical factors related to CIPR.
c. The overarching structure and inter-connectedness among the critical infrastructure domains.
d. The interaction among infrastructure systems under various degraded states of operation.
e. Possible conceptual and design solutions, and related information.
f. Strategies for verification and validation of solutions.

The CIPR WG will provide a collection of systems engineering and related products that provide understanding and solutions for domain stakeholders impacted by the events. This can include products developed by several working groups and initiatives, such as Architecture, Complex systems, Model-based systems engineering (MBSE), Decision analysis, Enterprise systems, Natural systems, Resilient systems, Risk management, Cost engineering, Human system interaction, In-service systems, Reliability engineering, Requirements, System of systems, System safety integration, Automotive, Healthcare, Infrastructure, Power & energy systems, Transportation systems, and Anti-terrorism. Other working groups also have knowledge to contribute as well. The CIPR WG will endeavor to integrate and coordinate among standards, regulations, and best practices of the impacted industries. It will also provide the organizing and development functions to establish new concepts and standards addressing CIPR.

Stakeholders with interest in CIPR are international and include all levels of government, defense and security agencies, critical infrastructure domain businesses and agencies, and society in general (e.g., regions, communities, and citizens).

3   SCOPE

Certain manmade and natural events have a known potential to affect societies at a national, continental, or even global scale. Such events can cause extreme harm well beyond those experienced from regional catastrophic events, especially when the effects will take longer than a month to recover. Three examples of events with the potential to cause critical infrastructure collapse include solar storms caused by coronal mass ejections (CME), electromagnetic pulse (EMP), and cyber events (intentional and otherwise). The CIPR WG will pursue its goals by addressing these three classes of events, and other classes of events with similar potential, when identified. The CIPR WG will promote and apply systems engineering principles with emphasis on policy, analysis, and concepts useful to understand, protect, and recover existing operational infrastructure, and to provide strategies, standards, and concepts for more resilient approaches. It will promote and perform activities supporting the stated goals.

This scope is synergistic with other INCOSE WGs identified above (e.g., MBSE, System of systems, Resilient systems, Power & energy, etc.). For example, the application of model-based approaches will be essential to analyze the problem and to communicate alternative conceptual solutions. Therefore, this WG will seek interest and participation from INCOSE members and the other INCOSE WGs. It will also reach out to engage international and governmental organizations, professional groups, critical infrastructure providers, and others stakeholders. MOUs, contracts, and other kinds of agreements may be sought with external organizations as needed to further the effort. These agreements, if any, will be established according to INCOSE guidelines, processes, and procedures.

The critical infrastructure domains addressed by the CIPR WG include the following. Other domains may be addressed as the need is identified:

1. Chemical and other industrial bases
2. Communications
3. Electrical & energy production and distribution
4. Emergency services
5. Financial services
6. Food and agriculture
7. Government services & facilities
8. Healthcare and public health
9. Information technology
10. Nuclear reactors, materials, and waste
11. Transportation
12. Water storage, treatment, and distribution
13. Waste handling and disposal (water, refuse, hazardous)
14. Society at large.

4   SKILLS AND EXPERTISE REQUIRED

CIPR WG needs members with the following skills and expertise. This list is not comprehensive and is not a requirement for membership.

- Organizational and technical leadership
- An advanced understanding of systems engineering
- Specialized engineering knowledge of critical infrastructure domains and associated systems and information
- Model-based methods and life-cycle management expertise
- An understanding of complex systems and systems of systems
- An understanding of system resilience and life-cycle operational availability
- A strong willingness to learn and advance the knowledge of technical/engineering perspectives of CIPR including organizational relationships and others perspectives as needed.

These skills will be drawn from INCOSE members, and from engagement with groups external to INCOSE.

5   MEMBERS, ROLES, AND RESPONSIBILITIES

Three co-chairs facilitate the group's decisions, which are made by consensus.

Role	Responsibilities
WG Lead chair	Be the primary POC for all WG activities, communications, and actions. This role includes relationships both internal and external to INCOSE. Responsible for annual budget and other financial activities. Responsible for operating process development and approval.
Technical co-chair	Convene monthly member meetings, maintain a list of technical projects and products, monitor progress on technical tasks, maintain the WG Connect site and external site, manage material and knowledge collection and distribution, and act as WG Lead when appropriate.
Logistics co-chair	Ensure that facilities and other resources are available for special meetings. Develop communications, programs, planning and arrangements for workshops, conferences, symposiums, and special public meetings.

6   OUTCOMES (PRODUCTS/SERVICES)

The following comprises the agenda for the next three years, and may change based on the needs of the systems engineering community. Products will support an international collection of stakeholders with interest in CIPR, including all levels of organizations and individuals from the list of domains in Section 3.
Products:

1. SOS, FOS, and System-specific systems engineering architecture products, documents, and other SE process products as needed.
   a. 2015: A list of targeted CIPR products being sought and developed, with anticipated completion dates. This list will be maintained annually.
   b. 2016: Select and generate initial products with emphasis on those that organize the critical infrastructure "Big Picture."
   c. 2017–2018: Develop specific targeted product from the list.
2. Pamphlets addressing CIPR topics from each relevant INCOSE working group

a. 2015: Put out a call to other INCOSE WGs for topics they are willing to contribute information to generate pamphlets or papers. This call will be conducted annually.

b. 2016–2018: Work with the other WGs to generate pamphlets and papers. Publish as they become ready.

3. Papers, articles, briefings, tutorials
   a. 2015: Call for papers in support of Energy Tech 2015 and IS2016. Calls will continue to be made in subsequent years.
   b. 2015: Development schedule for papers, articles, briefings, and tutorials based on proposals made in response to calls for these products. This schedule will be maintained annually.
   c. 2016: At least one paper and one briefing.
   d. 2017–2018: Papers, articles, briefings, and tutorials as scheduled.

4. Training courses taught by CIPR WG members
   a. 2016: A list of CIPR members who will develop and provide courses and tutorials.
   b. 2017: First course or tutorial on a CIPR topic taught by a CIPR member.
   c. 2018: Continued delivery of tutorials and courses based on requests and speaker availability. A schedule will be kept.

5. Provide references and other products to those needing information
   a. 2016: Make references available to members via INCOSE Connect. Maintain this site.
   b. 2017: Evaluate options for making key references available external to INCOSE.
   c. 2018: Establish and external communications presence that makes important resources available external to INCOSE.

Services:

1. Organize and support development of standards related to CIPR
   a. 2017: Establish a list of standards relevant to CIPR. Identify gaps in standards.
   b. 2018: Identify national and international policy issues related to CIPR.
   c. 2018: Work with a selected standards-generating body to address policy issues, or to initiate development of a new standard.

2. Support INCOSE conferences, meetings, etc. addressing CIPR domains
   a. 2015–2018: Hold WG meetings at INCOSE IW and IS.
   b. 2015: Support Energy Tech conference in conjunction with the INCOSE PESWG.
   c. 2016: Develop a plan for supporting development of future CIPR tracks at INCOSE conferences.

3. Support external organizations related to CIPR
   a. 2015–2016: Develop a list other organizations, external to INCOSE, related to CIPR that have the possibility of collaboration and outreach.
   b. 2016: Develop a strategy for outreach to the listed organizations.
   c. 2016–2018: Support external conferences, meeting, and other activities with the external organizations.
   d. 2016–2018: Pursue formalized relations with external organizations.

4. Provide expert assistance to users of CIPR WG products
   a. 2016: Develop a list of CIPR WG members able and willing to provide assistance to users of WG products.
   b. 2017–2018: Make the list available for use through INCOSE outreach processes.

5. Review results of CIPR-related research and developments (e.g., systems engineering products and courses) by systems engineers outside the CIPR WG.
   a. 2016–2018: Develop and maintain a list of products and courses developed by sources external to the CIPR WG, including other INCOSE WGs and groups external to INCOSE. This list will be developed by research of the members and made available through INCOSE Connect.

6. Capture lessons learned, Recommendations, etc. and apply to future products, research, etc.
   a. 2016–2018: Develop and maintain list of lessons learned and recommendations. The list will be updated annually and available to WG Members via INCOSE Connect.

7   APPROACH

- Meet monthly, on a day agreeable to a majority of the active group members. The purpose of the meeting is to share pertinent information, to review and advance technical projects, and to make decisions. Meetings will include call in, shared presentations, etc.
- Delegate between-meeting work on technical projects to one or more members and/or groups, who incorporate feedback from the group and flesh out details for discussion.
- Communicate by email and share results using INCOSE media, meetings, etc.
- Volunteers will be encouraged to participate in activities relevant to CIPR with organizations external to INCOSE, and to share resulting information and outcomes with the working group.
- Decisions regarding this charter will be made by the CIPR WG co-chairs as noted herein (see Section 5). Decisions regarding products, services, and other activities will be made by consensus among the participants active in the specific activity.
- INCOSE can help this working group reach members external to the membership by providing a web portal to disseminate information to interested parties, and may include items such as a website and a wiki. In addition, INCOSE can distribute a brochure during symposia and conferences that summarize the CIPR WG, the available products, and contact information.

8   MEASURES OF SUCCESS

- Papers & presentations submitted to the INCOSE Symposia, conferences, and other organizations.
- Papers published in INCOSE and other organizations' publications.
- Contributions of the group and individual members to efforts by other organizations.
- Number of members and guests at each CIPR WG meeting.
- Engagement of the WG and its members with external organizations related to CIPR.

9   RESOURCE REQUIREMENTS

CIPR WG will submit an annual budget request to obtain resources required to support this effort.
Working Group SharePoint Site on the INCOSE Connect.
Share Point Site external to INCOSE.
Need Global Meet account.
Resource requirements may require seeking resources external to INCOSE.

10   DURATION

The group will continue its efforts as long as there is a need to develop and communicate CIPR information and standards. This Charter will remain in effect until rescinded by the signatory.

11   SIGNATURES

WG Lead Chair	Date

Technical Director, INCOSE	Date

Revision History

Date	Revision	Description	Author
06/30/2015	0	Initial Issuance	CIPR Working Group

National Space Weather Strategy from websites beginning at this website:

http://www.dhs.gov/national-space-weather-strategy

National Space Weather Strategy

The technology and infrastructure that forms the backbone of America's economic vitality and national security is subject to many risks, and among the most challenging is the risk posed by space weather storms. In November 2014, the National Science and Technology Council established the Space Weather Operations, Research and Mitigation (SWORM) Task Force, and its charter directed the development of a National Space Weather Strategy, which will articulate high-level strategic goals for enhancing national preparedness to space weather events. Assistant Secretary for Infrastructure Protection Caitlin Durkovich is a co-chair of the SWORM Task Force.

Reducing the nation's vulnerability to space weather is a national priority. The Strategic National Risk Assessment identifies space weather as a hazard with the potential to pose a significant risk to national security. As a national risk, space weather warrants a coordinated strategy, and the Whole Community must work together to enhance the resilience of critical infrastructure to the potentially debilitating effects of space weather.

The SWORM Task Force recently released a draft national strategy and is seeking public comment. Everyone has a part to play in mitigating the nation's risk to space weather, and the Office of Infrastructure Protection encourages stakeholders to review the current draft.

Last Published Date: April 29, 2015

NATIONAL SPACE WEATHER STRATEGY

PRODUCT OF THE

National Science and Technology Council April 2015

Adapted from DHS website:
http://www.dhs.gov/sites/default/files/publications/DRAFT-NSWS-For-Public-Comment-508.pdf

Executive Summary

Reducing the Nation's vulnerability to space weather is a national priority. Space weather describes the variations in the space environment between the sun and Earth that can affect infrastructure systems and technologies in space and on Earth. It can disrupt the technology that forms the backbone of our economic vitality and national security, including satellite and airline operations, communications networks, navigation systems, and the electric power grid. These key components of our Nation's infrastructure and economy are increasingly at risk from space weather storms. The Strategic National Risk Assessment[1] identifies space weather as a hazard that poses significant risk to the security of the Nation.

This Strategy builds on recent significant efforts to reduce risks associated with natural hazards and improve the resilience of critical facilities and systems[2]. It aims to foster a collaborative environment in which government, industry, and private citizens can better understand and prepare for the effects of space weather. As a Nation, we must continue to leverage our existing national network of expertise and capabilities and pursue targeted enhancements to improve our ability to manage risks associated with space weather. With this Strategy, we seek to enhance the integration of existing national efforts and add important capabilities to help meet growing demands for space weather information. Six strategic goals underpin our efforts to reduce the Nation's vulnerability to space weather:

1. Establish benchmarks for space weather events: Effective and appropriate actions for space weather events require an understanding of the magnitude and frequency of storms. Benchmarks will help us assess the vulnerability of critical infrastructure and will provide critical points of reference to enable mitigation procedures and practices, as well as enhance response and recovery planning.

2. Enhance response and recovery capabilities: We must develop comprehensive guidance to support existing response and recovery constructs to manage space weather events with federal, state, local, tribal, and territorial governments and the private sector.

3. Improve protection and mitigation efforts: To build national preparedness we must improve our protection and mitigation efforts. Protection focuses on capabilities and actions to eliminate critical infrastructure vulnerabilities to space weather, and mitigation focuses on long-term vulnerability reduction and enhancing resilience to disasters. Together, these preparedness missions constitute our national effort to reduce the vulnerabilities and manage the risks associated with space weather events.

4. Improve assessment, modeling, and prediction of impacts on critical infrastructure: We must provide timely, actionable, and relevant decision-support services during space weather storms. Societal impacts

must also be understood to better inform the urgency of action during extreme events and to encourage appropriate mitigation and protection measures before an incident.

5. Improve space weather services through advancing understanding and forecasting: We must take action to improve the fundamental understanding of space weather. Accurate, reliable, and timely space weather observations and forecasts (and related products and services) are essential elements in enabling national preparedness. The underpinning science and observations that will help drive the necessary advances in modeling capability that supports user needs are the key to the quality of space weather products and services. We must also improve our capacity to develop and transition the latest scientific and technological advances into space weather operations centers.

6. Increase international cooperation: Because we live in a world of complex interdependencies, we need global engagement and a coordinated international response to space weather. We must not only be an integral part of the global effort, but must mobilize broad, global support. We will do so by utilizing existing agreements and by building international support at the policy level.

The National Strategy for Space Weather identifies national goals and establishes the guiding principles that will underpin our efforts to secure the critical technology infrastructures vital to our national security and economy. It identifies specific initiatives to drive both near- and long-term national protection priorities. It also provides protocols for preparing and responding to space weather events, ensuring that critical information is available to national leaders for informed decision making. This critical information will be used to enhance national resilience and prepare an appropriate response during space weather storms. This Strategy will facilitate the integration of space weather information into federal government risk-management plans to achieve desired levels of preparedness consistent with existing national policies.

Accomplishing the strategic elements in the Strategy will require a whole-of-government approach to coordinate and apply federal resources. It will also require us to strengthen public–private and international partnerships, using a Whole Community approach[3]. As a Nation, we must work together to enhance the resilience of critical infrastructure to the potentially debilitating effects of space weather, and we must ensure mechanisms are in place to help protect the people, economy, and national security of the United States.

Introduction

Space weather is a naturally occurring phenomenon that has the potential to negatively affect energy infrastructure, technology, and human health, which are essential contributors to national security and economic vitality. The term "space weather" refers to the dynamic conditions of the space environment that arise from interactions with emissions from the sun, including solar flares, solar energetic particles, and coronal mass ejections (CME). These emissions can affect Earth and its surrounding space, potentially causing disruption to electric power systems; satellite, aircraft, and spacecraft operations; telecommunications; position, navigation, timing (PNT) services; and other technology and infrastructure. Given the growing importance of reliable electric power and space-based assets for security and economic well-being, it is critical that we establish a strategy to improve the Nation's ability to protect, mitigate, respond to, and recover from the potentially devastating effects of space weather events.

Space weather is a global issue. Unlike terrestrial weather events (e.g., a hurricane), space weather has the potential to simultaneously affect the whole of North America or even wider geographic regions of the planet. The United States is currently a global leader in observing and forecasting space weather events, but our capability and situational awareness depend on international cooperation and coordination.

This Strategy outlines the objectives for enhancing the Nation's space weather readiness in three key areas: understanding, forecasting, and national preparedness. Federal departments and agencies have taken significant steps in these key areas. The challenges posed by the increasing vulnerability to space weather events require continuing research and development efforts to improve observation and forecast capabilities, which are linked directly to preparedness. This Strategy will leverage these efforts and existing policies while promoting enhanced coordination and cooperation across the public and private sectors in the United States and abroad.

Structure of the Strategy

This Strategy articulates six high-level goals for federal research, development, deployment, operations, coordination, and engagement:

1. establish benchmarks for space weather events;

2. enhance response and recovery capabilities;

3. improve protection and mitigation efforts;

4. improve assessment, modeling, and prediction of impacts on critical infrastructure;

5. improve space weather services through advancing understanding and forecasting;

6. increase international cooperation.

Implementation of the Strategy

The implementation of this Strategy will be overseen by the NSTC.

Enhancing National Preparedness and Critical Infrastructure Resilience

This Strategy ensures that space weather is fully integrated into the Presidential Policy Directive (PPD)-8, *National Preparedness* (March 30, 2011) and PPD-21, *Critical Infrastructure Security and Resilience* (February 12, 2013) frameworks. PPD-8 calls for an integrated, all-of-Nation, capabilities-based approach to preparedness for all hazards. The Directive also calls for the creation of a national planning framework. In support of this, the Department of Homeland Security coordinated the development of the Strategic National Risk Assessment (SNRA)[4]. The SNRA identified space weather as one of nine natural hazards with the potential to significantly affect homeland security.

PPD-21 identifies three strategic imperatives to drive the Federal approach to strengthen critical infrastructure security and resilience that are at the core of this Strategy[5]. The Directive identifies energy and communications systems as uniquely critical due to the enabling functions they provide across all critical infrastructure sectors. The Directive also instructs the federal government to engage with international partners to strengthen the security and resilience of domestic critical infrastructure and international critical infrastructure on which the Nation depends.

Strategic Goals

To meet the challenges presented by the negative effects of space weather events, this Strategy defines six strategic goals to prepare the Nation for near- and long-term space weather impacts. The objectives of these goals are to improve our understanding of, forecasting of, and preparedness for space weather events (phenomena and effects).

1. Establish Benchmarks for Space Weather Events

Developing benchmarks for space weather events is an important component to addressing the effects of space weather. Benchmarks are a set of characteristics and conditions against which a space weather event can be measured. They provide a point of reference from which to improve the understanding of space weather effects, develop more effective mitigation procedures, and enhance response and recovery planning. The objective of the benchmarks is to provide clear and consistent descriptions of the relevant physical parameters of space weather phenomena based on current scientific understanding and the historical record. For example, the benchmarks may serve as inputs to vulnerability assessments or defining points of action. But these benchmarks do not assign a category, classification, level, or significance of impact to an event. To be effective, the benchmarks must be developed in a timely manner using transparent methodology with a clear statement of assumptions and uncertainties. Because of relatively limited data and gaps in understanding space weather phenomena, benchmarks should be reevaluated as significant new data and research become available.

- o **Define scope, purpose, and approach for developing benchmarks:** Space weather benchmarks will be used to develop scenarios, inform practices (e.g., device, operational, and mitigation), and serve as reference points from which to develop impact and vulnerability assessments. The benchmarks will use multiple parameters to describe the space weather event. The parameters should include characteristics of the space weather event and the characteristics of its interactions with Earth and near-Earth environments (e.g., radio blackout and geomagnetic disturbance). Multiple benchmarks will be created to address:

    - o the different types of space weather events; for example, radio blackouts induced by solar flares and geomagnetic disturbances induced by CMEs;
    - o multiple physical parameters that will ensure the functionality of the benchmarks; for example, magnitude and duration; and
    - o a range of event magnitudes and associated recurrence intervals; for example, multiple event scenarios may inform different vulnerability thresholds, and an understanding of the "worst case" scenario may be instructive.

2. Enhance Response and Recovery Capabilities

Extreme space weather events are low-frequency, potentially high-impact events that will require a coordinated national response and recovery effort. Leveraging the National Planning Frameworks[6], the Nation will develop comprehensive guidance to support existing response and recovery constructs to manage extreme events with federal, state, local, tribal, territorial (SLTT), and other Whole Community partners[7]. Improving impact assessments and systems modeling will allow for greater planning fidelity for the effects of extreme events on critical infrastructure systems and the Whole Community. Likewise, improved forecasting capabilities enable development of time-sensitive procedures before any impacts. Building the Nation's restoration capability will require continued investments, unique solutions, and strong public–

private partnerships. The following objectives will be met to enhance response and recovery capabilities:

- **Complete an all-hazards power outage response and recovery plan:** The primary risk of an extreme space weather event is the potential for the long-term loss of electric power and the cascading affects that it would have on other critical infrastructure sectors; however, other low-frequency, high-impact events are also capable of causing long-term power outages on a regional or national scale. It is essential to have a comprehensive and executable plan (with key decision points) for regional or national power outages. The plan must include the Whole Community and enable the prioritization of core capabilities.

- **Support federal, SLTT government, and private sector planning for and managing of an extreme space weather event:** Information on the effects of an extreme space weather hazard on SLTT all-hazards planning is limited. Credible information and guidance on how to incorporate that knowledge into SLTT all-hazards planning will be developed and disseminated.

- **Provide guidance on contingency planning for extreme space weather impacts on the continuation of critical government and industry services:** A functional government, movement of personnel, preservation of services, and maintenance of critical infrastructure systems are essential before, during, and after an extreme space weather event. All levels of government, the private sector, and critical infrastructure entities will have guidance to respond in a manner that allows them to maintain the essential elements of their operations for a prolonged period of time.

- **Ensure communications systems capability and interoperability during extreme space weather events:** Effective communications systems are essential to gaining and maintaining situational awareness and ensuring unity of effort in response and recovery operations. While space weather affects communications systems, these effects can occur at different time scales within a single event and with varying impacts depending on the specific communications system, the characteristics of the event, and its duration. Government and private sector stakeholders must have guidance that allows them to maintain communication systems capabilities (including interoperability) during an extreme event.

- **Encourage the owners and operators of critical assets to coordinate the development of realistic power restoration priorities and expectations:** Electrical power providers should develop protocols for restoring electrical power before disruptions in coordination with State and local governments. Critical asset owners and operators must work with their providers to ensure that their power needs are understood. The owners and operators should consider plans and capabilities for temporary power in the event of an electrical power disruption caused by an extreme space weather event.

- **Develop and conduct exercises to improve and test federal, state, regional, local, and industry-related space weather response and recovery plans:** Evaluating the effectiveness of plans includes developing and executing a combination of training events and exercises to determine whether the goals, objectives, decisions, actions, and timing outlined in the plan support successful response and recovery. Exercising plans and capturing lessons learned enables ongoing improvement in event response and recovery capabilities.

- **Increase the Nation's restoration capability through continued investments, unique solutions, and strong public–private partnerships:** The Nation has not experienced the full consequences of an extreme space weather event in modern history. Improving the Nation's capability to respond to and recover from such an event will require continued investments and innovative solutions. Without strong public and private partnerships developed before such an event, however, an effective recovery will remain impractical.

3. Improve Protection and Mitigation Efforts

Growing interdependencies of critical infrastructure systems have increased the potential vulnerabilities to space weather events. Protection and mitigation efforts to eliminate or reduce space weather risks are essential missions of national preparedness. Protection focuses on capabilities and actions to eliminate critical infrastructure vulnerabilities to space weather, and mitigation focuses on long-term vulnerability reduction and enhancing the resilience to disasters[8]. Together, these preparedness missions constitute our national effort to reduce the vulnerabilities and manage the risks associated with space weather events. Four objectives are outlined for improving protection and mitigation efforts:

- **Assess the relevant legal mechanisms, authorities, and incentives that can be used to protect critical systems:** Statutory and regulatory authorities related to the protection of critical infrastructure already exist as do incentives for encouraging actions by critical infrastructure owners and operators. These will be identified along with the corresponding authorities, gaps, issues, and associated approaches to governance.

- **Encourage the development of hazard-mitigation plans that reduce vulnerabilities to, manage risks from, and assist with response to impacts associated with space weather:** In support of Whole Community planning for resilience, information about space weather hazards will be integrated, as appropriate, into existing mechanisms for information sharing, including Information Sharing Analysis Organizations, and into national preparedness mechanisms that promote strategic alignment between public and private sectors.

- **In concert with industry partners, achieve long-term vulnerability reduction to space weather events by implementing appropriate measures at critical locations most susceptible to space weather:** Adopting standards, business practices, and operational procedures that improve protection and resilience is essential to addressing space weather system vulnerabilities. The space weather benchmark events described in the first strategic goal (Establish Benchmarks for Space Weather Events) will be used to support the adoption of design standards for enhanced resilience; evaluate strategies for, priorities for, and feasibility of protecting critical assets; and foster mechanisms for sharing best practices that promote mitigation of damage from, and protection of, systems affected by space weather.

- **Strengthen public/private partnerships that support private action to reduce public vulnerability to space weather:** Private sector entities, as the owners and operators of the majority of the Nation's critical infrastructure, are essential to improving resilience. Space weather events do not respect national, jurisdictional, or corporate boundaries. Incorporating resilience measures into U.S. infrastructure systems requires collaboration, the support of existing coordinating mechanisms for information sharing and access, and identifying incentives and disincentives for investing in resilience measures.

4. Improve Assessment, Modeling, and Prediction of Impacts on Critical Infrastructure

A key component to improving national preparedness for a space weather event is the ability to observe and predict associated effects. Providing timely, actionable, and relevant decision-support services during a space weather event requires improvements in the ability to observe, assess, model, and ultimately predict the effects on critical national infrastructures such as the electric power systems, transportation systems, communications, and PNT systems. Societal impacts must also be understood to inform the urgency of action during events and to encourage appropriate mitigation and protection measures before an incident. Improving situational awareness and prediction of effects on infrastructure during an event requires better observations and better modeling of system-response characteristics. The following objectives will be met to enhance observation, modeling, and prediction capabilities:

- **Develop a national capability for real-time assessment of space weather impacts on critical systems:** Situational awareness of the state of various critical infrastructure systems is crucial to providing actionable event response. In addition, better and more thorough measurements of infrastructure responses to space weather events will inform and validate system-specific impact models that will ultimately improve event response. This capability will require continued investments in and assessments of the real-time monitoring requirements for reporting the state of infrastructures, as well as space weather situational awareness.

- **Develop or refine operational space weather impact/systems models:** It is not enough to forecast the magnitude of an impending geomagnetic disturbance for appropriate and effective response: it is also necessary to predict the effects of an event on infrastructure and other systems on a regional basis. Hurricane storm surge prediction is a terrestrial weather example of this objective. To do this effectively requires reliable, accurate, and fast models that take into account effects on both isolated and interdependent infrastructure systems. We must also define and develop comprehensive requirements for operational impact models, identifying deficiencies in current modeling capabilities to develop new and improved tools to achieve these objectives.

- **Improve operational impact forecasting and communications protocols:** Based on the assessment and modeling elements outlined above, a national capability to forecast extreme space weather effects before the onset of an event would enable timely warnings to system operators and emergency managers. This capability should always be available, with rapid computation and dissemination mechanisms.

- **Support basic and applied research into space weather impact on industries, operational environments, and infrastructure sectors:** Improving existing models and developing new capabilities in impact forecasting must be based on a better understanding of the fundamental physical processes of space weather impacts to critical infrastructure systems. Doing so requires identifying gaps in our understanding of impacts on critical national infrastructures; developing strategies to address these gaps; identifying impact-related interdependencies through vulnerability and failure mode-assessments across and between sectors; and supporting research for understanding the cost required to mitigate, respond to, and recover from an extreme space weather event.

5. Improve Space Weather Services through Advancing Understanding and Forecasting

Space weather services can enhance national preparedness by providing timely, accurate, and relevant forecasting products. Identifying and sustaining a baseline of critical measurements from observing platforms is key to providing operational services that inform preparedness. This baseline can also serve as a reference point from which to identify coverage and measurement gaps, as well as opportunities for improvement. Services can be improved through basic and applied research that focuses on the needs of an increasingly diverse user community. To facilitate the transition of these enhancements from the research domain to operations, the responsible agencies will (1) periodically revalidate user requirements for improved space weather services and (2) strengthen and encourage partnerships to accelerate the research-to-operations transition process, with a goal to support key preparedness decisions. Seven objectives are outlined to meet these goals:

- **Define a baseline operational space weather observation capability:** Our Nation currently lacks a comprehensive operational space weather observation strategy. Although operational systems are robust, resilient, and ensure the data continuity necessary for a national space weather prediction capability exists, currently, an ad hoc mixture of weather and research satellites and ground systems is being used to provide critical data to forecast centers. To ensure adequate and sustained real-time observations for space weather analysis, forecasting, and decision-support services, a baseline, or minimally adequate, operational observation capability should be defined. The observation baseline will also specify the optimal mix of ground-based and satellite observations to enable 320 continuous and timely space weather watch, warning, and alert products and services.
- **Improve understanding of user needs for space weather forecasting and use these data to establish lead-time and accuracy goals:** Effective transfer of space weather knowledge requires a better understanding of the effects of space weather on technology and on industry and government customers, including the associated economic and political impacts on the Nation's critical infrastructures.

- **Ensure products are intelligible and actionable to inform critical decision making:** Decision-relevant information must be communicated in ways that stakeholders can fully understand and use. Models and forecasts must enable swift decision making with a reasonable assumption of risk.

- **Improve forecasting accuracy and lead time:** Society is increasingly at risk to extreme space weather events. With improved predictions, our Nation can enhance mitigation, response, and recovery actions to safeguard our assets and maintain continuity of operations during high-impact space weather activity.

- **Enhance fundamental understanding of space weather and its drivers to develop and continually improve predictive models:** Forecasting space weather depends on a fundamental understanding of the space environment processes that give rise to hazardous events. Particularly important is understanding the processes that link the sun to Earth. An improved understanding will help drive the necessary advances in modeling capability to support user needs.

- **Improve effectiveness and timeliness of research to operations transition process:** Although the Nation has invested in the development of research infrastructure and predictive models to meet the demands of a growing space weather user community, existing modeling capabilities still fall short of providing what is needed to meet these critical demands. Until better research models are incorporated into operational forecasts, the Nation will not fully realize the benefits of its research investments.

- **Assess and develop observational strategies for the study and prediction of space weather events:** Fundamental research, modeling, product development, and space weather hazard assessments require observations taken from space and on the ground. Development of advanced technologies has the potential to improve the quality and affordability of new observing systems and optimize the path from research to operational use. Coordination between the space weather research and operations communities to identify critical observational data products required to advance predictive modeling capability is necessary to sustain critical space weather observing capabilities. It is also important to explore the needs for improved coverage, timeliness, and data quality through partnerships with academia, the private sector, and international collaborators.

6. Increase International Cooperation

In a world increasingly dependent on interconnected and interdependent infrastructure, any disruption to these critical technologies could have regional and even international consequences. Therefore, space weather should be regarded as a global challenge requiring a coordinated global response. Many countries are becoming increasingly aware of the need to monitor and manage space weather risks. The United States and other nations have begun sharing observations and research, disseminating products and services, and collaborating on real-time predictions to mitigate impacts on critical technology and infrastructure. We must work together to foster global collaboration, taking advantage of mutual interests and capabilities to improve situational awareness, predictions, and preparedness for extreme space weather.

The following objectives will be met to increase international cooperation:

- **Build international support at the policy level for space weather as a global challenge:** A prerequisite to enhanced international cooperation is high-level support across partner countries to raise awareness of space weather as a global challenge.

- **Promote a collaborative international approach to protect against, mitigate the effects of, respond to, and recover from extreme space weather events:** The world's interconnected and interdependent systems are vulnerable to extreme space weather events; this vulnerability could possibly lead to a cascade of impacts across borders and sectors. To mitigate these risks, we will work with the international community to facilitate the exchange of information and best practice to strengthen global preparedness capacity for extreme space weather events. We will also foster the development of global mutual aid agreements to facilitate response and recovery efforts and coordinate international partnership activities to support space weather preparedness and response exercises.

- **Increase engagement with the international community on scientific research, observation infrastructure, and modeling:** Gaps in research, observations, models, and forecasting tools need to be identified and filled to meet the needs of the global scientific community and the providers and users of space weather information services.

- **Improve international data sharing:** Increased access to government, civilian, and commercial space weather data across the globe is of mutual benefit to the United States and partner nations.

- **Strengthen international coordination and cooperation on space weather products and services:** Providing high-quality space weather products and services worldwide requires international coordination and cooperation. Toward this end, we will seek agreement on common terminology, measurements, and scales of magnitude; promote and coordinate the sharing and dissemination of space weather observations, model outputs, and forecasts; and establish coordination procedures across space weather operations centers during extreme events.

- **Develop coherent international communication strategies:** The global hazards of space weather must be clearly communicated to policymakers, the public, and the technical community. A process is needed to (1) issue forecasts, alerts, and warnings using consistent nomenclature and nontechnical terminology where appropriate and (2) promote and support public outreach and space weather education globally.

Conclusion

Space-weather events pose a significant and complex risk to the Nation's infrastructure and have the potential to cause substantial economic and human harm. This Strategy is the first step in addressing the myriad challenges presented when managing and mitigating the risks posed by both severe and ordinary space weather. As outlined above, the six high-level goals and their associated objectives support a collaborative and federally coordinated approach to developing effective policies, practices, and procedures for decreasing our Nation's vulnerabilities. By establishing goals for improvements in forecasting, research, preparedness, planning, and domestic and international engagement, this Strategy will help ensure our Nation's ability to withstand and quickly recover from effects of extreme space weather events.

References

National Space Policy (June 28, 2010)

National Strategy for Civil Earth Observations (April 2013)

Presidential Policy Directive 8 (PPD-8): National Preparedness (March 30, 2011)

Presidential Policy Directive 21 (PPD-21): Critical Infrastructure Security and Resilience (February 12, 2013)

The Strategic National Risk Assessment in Support of PPD 8: A Comprehensive Risk-Based Approach Toward a Secure and Resilient Nation (December 2011)

The National Aeronautics and Space Administration Authorization Act of 2010 (October 11, 2010)

Abbreviations

CME coronal mass ejection

NSTC National Science and Technology Council

OSTP Office of Science and Technology Policy

PNT position, navigation, timing

PPD Presidential Policy Directive

R&D research and development

SLTT State, local, tribal, territorial

SNRA Strategic National Risk Assessment

Appendix: Background on Solar Phenomena that Drive Space Weather

Space weather is commonly driven by solar storm phenomena that include coronal mass ejections (CMEs), solar flares, solar particle events, and solar wind. These phenomena can occur anywhere on the sun's surface, but only solar storms that are Earth directed are the potential drivers of space weather events on Earth. An understanding of solar storm phenomena is an important component to developing accurate space weather forecasts (event onset, duration, and magnitude). CMEs are explosions of plasma (charged particles) from the sun's corona. They generally take two to three days to arrive at Earth, but in the most extreme cases, have been observed to arrive in as little as 17 hours. When CMEs collide with Earth's magnetic field, they can cause a space weather event called a geomagnetic storm, which often includes enhanced auroral displays. Geomagnetic storms of varying magnitudes can cause significant long- and short-term impacts to the Nation's critical infrastructure, including the electric power grid, aviation systems, GPS applications, and satellites. A solar flare is a brief eruption of intense high-energy electromagnetic radiation from the sun's surface, typically associated with sunspots. Solar flares can affect Earth's upper atmosphere, potentially causing disruption, degradation, or blackout of satellite operations, radar, and high-frequency radio communications. The electromagnetic radiation from the flare takes approximately eight minutes to reach Earth, and the effects usually last for one to three hours on the daylight side of Earth.

Solar particle events are injections of energetic electrons, protons, alpha particles, and other heavier particles into interplanetary space. Following an event on the sun, the fastest moving particles can reach Earth within tens of minutes and temporarily enhance the radiation level in interplanetary and near-Earth space. When energetic protons collide with satellites or humans in space, they can penetrate deep into the object that they collide with and cause damage to electronic circuits or biological DNA. Solar particle events can also pose a risk to passengers and crew in aircraft at high latitudes near the geomagnetic poles and can make radio communications difficult or nearly impossible.

Solar wind, consisting of plasma, continuously flows from the sun. Different regions of the sun produce winds of different speeds and densities. Solar wind speed and density play an important role in space weather. High-speed winds tend to produce geomagnetic disturbances while slow-speed winds can bring calm space weather. Space weather effects on Earth are highly dependent on solar wind speed, solar wind density, and direction of the magnetic field embedded in the solar wind. When high-speed solar wind overtakes slow-speed wind or when the magnetic field of solar wind switches polarity, geomagnetic disturbances can result.

National Science and Technology Council

The National Science and Technology Council (NSTC) was established by Executive Order on November 23, 1993. This Cabinet-level Council is the principal means within the executive branch to coordinate science and technology policy across the diverse entities that make up the federal research and development enterprise. Chaired by the President, the membership of the NSTC is made up of the Vice President, the Director of the Office of Science and Technology Policy, Cabinet Secretaries and Agency Heads with significant science and technology responsibilities, and other White House officials.

A primary objective of the NSTC is the establishment of clear national goals for federal science and technology investments in a broad array of areas spanning virtually all the mission areas of the executive branch. The Council prepares research and development strategies that are coordinated across federal agencies to form investment packages aimed at accomplishing multiple national goals. The work of the NSTC is organized under five primary committees: Environment, Natural Resources and Sustainability; Homeland and National Security; Science, Technology, Engineering, and Math (STEM) Education; Science; and Technology. Each of these committees oversees subcommittees and working groups focused on different aspects of science and technology and working to coordinate across the federal government.

For additional information concerning the work of the National Science and Technology Council please send an email to: nstc@ostp.gov.

Endnotes

[1] *The Strategic National Risk Assessment (SNRA) in Support of PPD 8: A Comprehensive Risk-Based Approach toward a Secure and Resilient Nation,* Department of Homeland Security, December 2011.

[2] See reference section for listing of recent relevant policy documents.

[3] FEM!, "A Whole Community Approach to Emergency Management: Principles, Themes, and Pathways for Action," FDOC 104-008-1, Department of Homeland Security, December 2011.

[4] *The Strategic National Risk Assessment in Support of PPD 8: A Comprehensive Risk-Based Approach toward a Secure and Resilient Nation,* Department of Homeland Security, December 2011.

[5] (1) Refine and clarify functional relationships across the federal government to advance the national unity of effort to strengthen critical infrastructure security and resilience; (2) enable effective information exchange by identifying baseline data and systems requirements for the federal government; and (3) implement an integration and analysis function to inform planning and operations decisions regarding critical infrastructure.

[6] The National Planning Framework describe how the Whole Community works together to achieve the National Preparedness Goal of "a secure and resilient nation with the capabilities required across the whole community to prevent, protect against, mitigate, respond to, and recover from the threats and hazards that pose the great risk." This Goal is the cornerstone for the implementation of PPD-8 (https://www.fema.gov/national-planning-frameworks).

[7] Whole Community planning for resilience is an approach to emergency management that reinforces the ideas that FEMA is only one part of our Nation's emergency management team—that we must leverage all the resources of our collective team in preparing for, protecting against, responding to, recovering from, and mitigating against all hazards; and that collectively we must meet the needs of the entire community in each of these areas (https://www.fema.gov/whole-community).

[8] Disaster resilience refers to the capability to prevent, or protect infrastructure from, significant multi-hazard threats and incidents and to expeditiously recover and reconstitute critical services with minimum damage to public safety and health, the economy, and national security (https://training.fema.gov/hiedu/docs/terms%20and%20definitions/terms%20and%20definitions.pdf).

Island-mode Enhancement Strategies and Methodologies for Defense Critical Infrastructure

Defense Threat Reduction Agency (DTRA) 2015.2—Topic DTRA152-006
Opens: May 26, 2015—Closes: June 24, 2015

DTRA152-006 TITLE: Island-mode Enhancement Strategies and Methodologies for Defense Critical Infrastructure

TECHNOLOGY AREAS: Nuclear Technology

The technology within this topic is restricted under the International Traffic in Arms Regulation (ITAR), which controls the export and import of defense-related material and services. Offerors must disclose any proposed use of foreign nationals, their country of origin, and what tasks each would accomplish in the statement of work in accordance with section 5.4.c.(8) of the solicitation.

OBJECTIVE: Develop and test innovative strategies and methodologies to enhance island-mode technologies and innovative tactics for defense critical infrastructure (DCI) in the event of commercial power grid loss/disruption due to an electromagnetic pulse (EMP) or high power microwaves (HPM).

DESCRIPTION: The defense critical infrastructure (DCI) is the composite of DoD and non-DoD assets essential to project, support, and sustain military forces and operations worldwide. The DCI includes, but is not limited to, elements such as military bases, ballistic missile defense installations, radar sites, etc. An electromagnetic (EM) attack (nuclear electromagnetic pulse [EMP] or non-nuclear EMP [e.g., high-power microwave, HPM]) has the potential to degrade or shut down portions of the electric power grid important to the DoD. While a power grid may employ intentional islanding techniques to protect sections of the grid and prevent a cascading collapse of the power grid, the broad reach of potential EM attacks with the potential of simultaneous levels of disruption might prevent traditional islanding protection methods from being sufficient for continued operations of the DCI. Restoring the commercial grid from the still functioning regions may not be possible or could take weeks or months. Significant elements of the DCI require uninterrupted power for prolonged periods to perform time-critical missions (e.g., sites hardened to MIL-STD-188-125-1). To ensure these continued operations, DCI sites must be able to function as a microgrid that can operate in both grid-connected and intentional island-mode (grid-isolated). Such a microgrid is defined as a group of interconnected loads and distributed energy resources within clearly defined electrical boundaries that acts as a single controllable entity with respect to the power grid. The purpose of this topic is, through systematic study of a typical DCI site, to develop enhanced methodologies and technologies for providing intentional island-mode capability at DCI sites in the event of grid loss. Methodologies should account for the need of immediate and continuous operations at sites and the seamless transition to and from commercial power (grid-connected and grid isolated states). The emphasis of this project should be on determining how to best prepare an existing DoD site for intentional island-mode operation and identifying major risks and hurdles. This work will require refinement of existing technologies and development of new technologies and is directed specifically toward applying the new knowledge to meet the survivability of DoD sites to EM attacks affecting large geographical areas. The goal of this project is to develop a set of methodologies and strategies that can be applied, along with existing methods, to enhance the resilience of DCI assets such as military bases. Such methods should aide in the development of islanding at DoD sites to ensure survivability to geographically large EM threats. These methods may also be applied to the commercial sector and other areas of the government: hospitals, civilian infrastructure, businesses, etc.

PHASE I: The successful Phase I project should develop innovative strategies and methodologies for DCI island-mode operations in the event of power grid disruption or failure due to an EM threat.

Sufficient detail should be developed to show technical competency and/or proof of concept. Phase I deliverables should also include a draft test plan detailing a testing approach to demonstrate these strategies as well as establishing performance goals. Additionally, a draft roadmap should be developed indicating Phase II and Phase III plans, timelines, and addressing key decision points and milestones.

PHASE II: Phase II will focus on intentional island-mode methodologies and strategies at a specific DCI site (TBD). Limited initial testing may occur at a proto-type site, via modeling, or prior to full scale testing at a DCI site. Identify and address key island-mode hurdles, limitations, and obstacles and provide recommendations on addressing these areas. Methodologies and strategies should be improved and expanded based on testing, assessments, and available date. Clear documentation on strategies/methodologies and improvements is a priority. Identification of dual use commercial applications is an important aspect of this phase.

PHASE III: The Phase III project would focus execution of the Phase II test plan and on expanding these methodologies and strategies to include systems/infrastructure outside the DCI. This could include other DoD/government agency sites, hospitals, civilian infrastructure, or other commercial sites. Methodologies developed for the site-specific work in Phase II could be expanded for a different site or generalized to create overarching guidelines.

REFERENCES:

1. DoDI DoD 3020.40. 2008. "Policy and Responsibilities for Critical Infrastructure," July 1, 2010. "Report of the Commission to Assess the Threat to the United States from Electromagnetic Pulse (EMP) Attack." Critical National Infrastructures. http://www.empcommission.org/docs/A2473-EMP_Commission-7MB.pdf.

2. "Report of the Commission to Assess the Threat to the United States from Electromagnetic Pulse (EMP) Attach." Volume 1: Executive Report, 2004. http://www.empcommission.org/docs/empc_exec_rpt.pdf.

CLEAN POWER PLAN AND ELECTRIC RELIABILITY

Felder, Frank A., "Should Relief Be Granted from the Clean Power Plan for Reliability reasons?" Electricity Journal 28:6; 5-11, July 2015.

North American Electric Reliability Corporation, Potential Reliability Impacts of EPA's Proposed Clean Power Plan: Initial Reliability Review, November 2014, ii + 27 pp.
http://www.nerc.com/pa/RAPA/ra/Reliability%20Assessments%20DL/Potential_Reliability_Impacts_of_EPA_Proposed_CPP_Final.pdf

North American Electric Reliability Corporation, Potential Reliability Impacts of EPA's Proposed Clean Power Plan, Phase I, April 2015. Ix + 59 pp. http://www.nerc.com/pa/RAPA/ra/Reliability Assessments DL/Potential Reliability Impacts of EPA%E2%80%99s Proposed Clean Power Plan - Phase I.pdf

Regional Greenhouse Gas Initiative, Inc., RGGI Joint Comments, EPA Docket EPA-HQ-OAR-2013-0602, filed November 5, 2014. 31 pp. http://www.rggi.org/docs/PressReleases/PR110714_CPP_Joint_Comments.pdf

U.S. Environmental Protection Agency, Clean Power Plan Final Rule, August 3, 2015. 1560 pp.
http://www.rggi.org/docs/PressReleases/PR110714_CPP_Joint_Comments.pdf

ECONOMICS OF RESILIENCY

Anderson, G. B., and Bell, M. L., "Lights out: Impact of the August 2003 power outage on mortality in New York, NY" Epidemiology 2012 March ; 23(2): 189–193. doi:10.1097/EDE.0b013e318245c61c.
http://www.ncbi.nlm.nih.gov/pmc/articles/PMC3276729/pdf/nihms348988.pdf

Bate, Jon D., Preliminary Economics Analysis of Electric Grid Protection Against Geomagnetic Disturbance (GMD) Events, Appendix 1, pp. 71-76, in Foundation for Resilient Societies, Comments on Reliability Standard for Transmission System Planned Performance for Geomagnetic Disturbance Events FERC Docket RM15-11-000, filed August 10, 2015.
http://www.resilientsocieties.org/uploads/5/4/0/0/54008795/docket_rm15-11-000_resilient_societies_corrected_20150810.pdf

Dobbins, Robert W. and Schrijver, Karel, Electrical Claims and Space Weather: Measuring the visible effects of an invisible force Zurich: Zurich Insurance Group, June 2015, 16 pp.
http://www.zurichservices.com/ZSC/REEL.nsf/9f359b3938a6bdd448257a4f001c4596/be82cb09b6294a6286257e7e00549d61/$FILE/wp_ElectricalClaims_SpaceWeather.pdf

Lloyd's, and the Centre for Risk Studies, University of Cambridge, Business Blackout: the insurance implications of a cyber attack on the US power grid Cambridge: Centre for Risk Studies, May 2015,
65 pp. https://www.lloyds.com/~/media/files/news and insight/risk insight/2015/business blackout/business blackout20150708.pdf

Lloyd's, and Atmospheric and Environmental Research, Solar Storm Risks to the North American Electric Grid June 2013. 21 pp. https://www.lloyds.com/~/media/lloyds/reports/emerging risk reports/solar storm risk to the north american electric grid.pdf

Morrall, John F., "Saving Lives: A Review of the Record," Working Paper, AEI-Brookings Joint Center for Regulatory Studies, 2003, and Journal of Risk and Uncertainty, 27:3; 221-237, 2003. http://biotech.law.lsu.edu/cases/adlaw/cba/Morrall-Saving-Lives-2003.pdf

Popik, Thomas S., et al., Electromagnetic Pulse and Geomagnetidc Disturbance Mitigation Cost Estimates_Rev18, Nashua, NH: Foundation for Resilient Societies, 2015. http://www.resilientsocieties.org/images/EMP_and_GMD_Cost_Estimate_Rev18_Public.xls

Schrijver, C. J., R. Dobbins, W. Murtagh, S. M. Petrinec, "Assessing the impact of space weather on the electric power grid based on insurance claims for industrial electrical equipment," Space Weather doi:10.0112/2014SW001066, 487-498, July 8 2014. http://onlinelibrary.wiley.com/doi/10.1002/2014SW001066/pdf

EMP RESOURCES

Baker, George H. Nuclear EMP Hardening Approach as the Basis for Unified Electromagnetic Environmental Effects Protection, James Madison University, Harrisonburg, VA, January 1992. http://works.bepress.com/george_h_baker/43

IEEE, IEEE 1642:2015 Recommended Practice For Protecting Publicly Accessible Computer Systems From Intentional Electromagnetic Interference (Iemi). http://shop.standards.ie/nsai/details.aspx?ProductID=1792841

Manto, Charles "New Federal Initiatives Planning for Long-term Power Grid Outages", radio interview 2015 available on the web here: http://www.offthegridnews.com/?powerpress_pinw=58937-podcast

Radasky, William A. "Electromagnetic Warfare is Here – A briefcase-size radio weapon could wreak havoc in our networked world"; 25 August 2014, Spectrum, IEEE, http://spectrum.ieee.org/aerospace/military/electromagnetic-warfare-is-here

NATIONAL SPACE WEATHER STRATEGY

National Science and Technology Council, National Space Weather Strategy, Draft Report for Public Comment, April 29, 2015. iii + 13 pp. http://www.dhs.gov/sites/default/files/publications/DRAFT-NSWS-For-Public-Comment-508.pdf

United Kingdom, Cabinet Office, Space Weather Preparedness Strategy, Version 2.1, July 2015. 40 pp. https://www.gov.uk/government/uploads/system/uploads/attachment_data/file/449593/BIS-15-457-space-weather-preparedness-strategy.pdf

RELIABILITY STANDARDS FOR CYBER SECURITY

Campbell, Richard J.Cybersecurity Issues for the Bulk Power System, Congressional Research Service, June 10, 2015. Iv + 35 pp. https://www.fas.org/sgp/crs/misc/R43989.pdf

CERT-UK, Cyber-security risks in the supply chain, CERT-UK, Feb. 2015, 11 pp. https://www.cert.gov.uk/wp-content/uploads/2015/02/Cyber-security-risks-in-the-supply-chain.pdf

Cotter, George R., et al., Comments on Reliability Standards for Cybersecurity and Supply Chain Management, FERC Docket RM15-14-000, Foundation for Resilient Societies, September 21, 2015.

Federal Energy Regulatory Commission, Notice of Proposed Rulemaking: Revised Critical Infrastructure Protection Reliability Standards. FERC Docket RM15-14-000, 152 FERC ¶ 61,054. July 16, 2015. Ii + 57 pp. at 80 Federal Register 43354-43367, July 22, 2015. http://www.gpo.gov/fdsys/pkg/FR-2015-07-22/pdf/2015-17920.pdf

Rogers, Admiral Mike S. Testimony, Public Hearing, "Cybersecurity Threats: The Way Forward," U.S. House Permanent Select Committee on Intelligence, Transcript, Federal News Service, November 20, 2014.https://www.nsa.gov/public_info/_files/speeches_testimonies/ADM.ROGERS.Hill.20.Nov.pdf

Swartz, Scott D., and Michael J. Assante, Industrial Control System (ICS) Cybersecurity Response to Physical Breaches of Unmanned Critical Infrastructure Sites SANS Institute, January 1, 2014. https://www.sans.org/reading-room/whitepapers/analyst/industrial-control-system-ics-cybersecurity-response-physical-breaches-unmanned-critical-in-35282

RELIABILITY STANDARD FOR PHYSICAL SECURITY

Federal Energy Regulatory Commission, Notice of Proposed Rulemaking re Physical Security Reliability Standard, RM14-15-000, July 17, 2014, 146 FERC ¶ 61,140

Federal Energy Regulatory Commission, Physical Security Reliability Standard, Docket RM14-15-000; Order No. 802, issued November 20, 2014, 149 FERC ¶ 61,166, 79 Federal Register 70069. https://www.ferc.gov/whats-new/comm-meet/2014/112014/E-4.pdf

Foundation for Resilient Societies, Reply Comments on Physical Security Standards, Sep. 22, 2014. http://resilientsocieties.org/images/RM14-15-000_Resilient_Societies_Sept_22_2014.pdf

Foundation for Resilient Societies, Request for Rehearing of FERC Order No. 802.Request for Rehearing of FERC Order No. 802 and remand of reliability Standard for Physical Security, Dec. 21, 2014. 20141222-5384(30011917).pdf http://www.ferc.20141222-5384(30011917)

Federal Energy Regulatory Commission, Order Denying Request for Rehearing, RM14-15-001, April 23, 2015, 151 FERC ¶ 61,066, http://www.ferc.gov/CalendarFiles/20150423170729-RM14-15-001.pdf

RELIABILITY STANDARDS FOR SOLAR GEOMAGNETIC DISTURBANCES – OPERATIONS

Federal Energy Regulatory Commission, Order No. 779, Final Rule Geomagnetic Disturbances RM12-22-000, issued May 16, 2013, 143 FERC ¶ 61147. http://www.ferc.gov/whats-new/comm-meet/2013/051613/E-5.pdf

Federal Energy Regulatory Commission, Order No. 797, Reliability Standard for Geomagnetic Disturbance Operations, Docket RM14-1-000, June 19, 2014, 147 FERC ¶ 61.209. http://www.ferc.gov/whats-new/comm-meet/2014/061914/E-18.pdf

Foundation for Resilient Societies, Supplemental Information Supporting Rehearing of Order No. 797 - GIC Monitoring Networks, FERC Docket RM14-1-001, Filed July 21, 2014. http://www.resilientsocieties.org/images/Resilient_Societies_Additional_Facts081814.pdf

Foundation for Resilient Societies, Supplemental Information Supporting Request for Rehearing - GIC Monitors, FERC Docket RM14-1-0001, filed August 18, 2014. http://www.resilientsocieties.org/images/Resilient_Societies_Additional_Facts081814.pdf

Federal Energy Regulatory Commission, Order No. 797-A, Denying Rehearing, Reliability Standard EOP-010-1, Geomagnetic Disturbance Operations, Docket No. RM14-1-001, October 16, 2014, 149 FERC ¶ 61,027. http://www.ferc.gov/whats-new/comm-meet/2014/101614/E-4.pdf

RELIABILITY STANDARDS FOR SOLAR GEOMAGNETIC DISTURBANCES – ASSESSMENT OF CRITICAL HARDWARE PROECTION

North American Electric Reliability Corporation, Level 2 Appeal, Foundation for Resilient Societies, Inc. Project 2013-03 Geomagnetic Disturbance Mitigation, including record of Appeal, March 2015.
http://www.nerc.com/pa/Stand/Project201303GeomagneticDisturbanceMitigation/2013-03_GMD_Level_2_Appeal_Foundation_for_Resilient_Societies_TPL-007-1_05182015.pdf

Federal Energy Regulatory Commission, Notice of Proposed Rulemaking, Reliability Standard for Transmission System Planned Performance for Geomagnetic Disturbance Events , Docket RM15-11-000, May 14, 2015, 151 FERC ¶ 61,134.
http://www.ferc.gov/whats-new/comm-meet/2015/051415/E-1.pdf

North American Electric Reliability Corporation, Transcript of NERC Level 2 Appeal of Procedures & Quality Control for Geomagnetic Disturbance Standard, June 29, 2015, August 17, 2015. 28 pp.
http://www.nerc.com/pa/stand/project201303geomagneticdisturbancemitigation/final_transcript_of_06292015_level_2_appeal_of_foundation_for_resilient_societies.pdf

North American Electric Reliability Corporation, Subcommittee of the NERC Board of Trustees, Level 2 Appeals Decision Proposed Reliability Standard TPL-007-1 (Project 2013-03 Geomagnetic Disturbance Mitigation), August 17, 2015, ii+13pp.
http://www.nerc.com/pa/stand/project201303geomagneticdisturbancemitigation/final_l2_panel_decision_foundation_for_resilient_societies_08172015.pdf

Foundation for Resilient Societies, Corrected Comments on Reliability Standard TPL-007-1 (Standard for Geomagnetic Disturbance Hardware Assessment), filed July 27, 2015, as Corrected, August 10, 2015. 91 pp.
http://www.resilientsocieties.org/uploads/5/4/0/0/54008795/docket_rm15-11-000_resilient_societies_corrected_20150810.pdf

STATE INITIATIVES

Andrea Boland, "Maine Legislation on Power Grid Protection from Manmade and Natural EMP"; Mitigating High-Impact Threats to Critical Infrastructure; pp. 130-140, Westphalia Press, Washington, DC 2014.
http://westphaliapress.org/2014/11/26/mitigating-high-impact-threats-to-critical-infrastructure-2013-conference-proceedings-of-the-infragard-national-emp-sig-sessions-at-the-dupont-summit/

Central Maine Power, 2014 Maine GMD/EMP Impacts assessment, December 23, 2014. Lookup Docket 2013-00415. Download item 51. https://www.mpuc-cms.maine.gov

Emprimus, Report to the Maine Public Utilities Commission, Report: Effects of GMD & EMP on the State of Maine Power Grid, Jan.2, corrected Jan. 5, 2015. 46 pp. http://www.maine.gov/tools/whatsnew/attach.php?id=639058&an=2